HOW TO FALL IN LOVE WITH QUESTIONS

HOW TO FALL IN LOVE WITH QUESTIONS

A NEW WAY TO THRIVE IN TIMES OF UNCERTAINTY

ELIZABETH WEINGARTEN

An Imprint of HarperCollinsPublishers

The names and identifying characteristics of some of the individuals featured throughout this book have been changed to protect their privacy.

HOW TO FALL IN LOVE WITH QUESTIONS. Copyright © 2025 by Elizabeth Weingarten. All rights reserved. Printed in the United States of America. No part of this book may be used or reproduced in any manner whatsoever without written permission except in the case of brief quotations embodied in critical articles and reviews. For information, address HarperCollins Publishers, 195 Broadway, New York, NY 10007.

HarperCollins books may be purchased for educational, business, or sales promotional use. For information, please email the Special Markets Department at SPsales@harpercollins.com.

FIRST EDITION

Designed by Elina Cohen

Library of Congress Cataloging-in-Publication Data has been applied for.

ISBN 978-0-06-333513-4

25 26 27 28 29 LBC 5 4 3 2 1

TO MY MOM, DAD, HUSBAND, AND SON—

my love for all of you is never in question.

CONTENTS

Introduction 1

PART ONE: DON'T LOOK FOR THE ANSWERS NOW

CHAPTER 1: The Charlatans of Certainty 15

CHAPTER 2: The Death (and Rebirth) of Patience 45

CHAPTER 3: Seeking the Right Sources to Ask 71

PART TWO: LOVE THE QUESTIONS THEMSELVES

CHAPTER 4: What Does a Life of Loving Questions Look Like? 97

CHAPTER 5: How Do We Fall in Love with Questions? 133

CHAPTER 6: Converting Questions into Freedom 169

BRIDGE TO PART THREE: Making Your Questions Map 189

PART THREE: LIVE THE QUESTIONS

CHAPTER 7: Reviving Curiosity When Questions Are Painful   199

CHAPTER 8: Learning How to Talk to Yourself   225

CHAPTER 9: Holding On and Letting Go   245

EPILOGUE: Where to Find Courage?   267

Acknowledgments   275

Appendix   277

Notes   281

INTRODUCTION

It was one of those nights I barely slept. Lying in bed in late June, next to my oblivious husband, I felt my heart pounding in my ears, my chest tight with anxiety.

If I did manage to sleep briefly, it was not the kind of sleep that helps you sort through a potentially disastrous life decision. It was certainly not the kind that people talk about when they tell you to "sleep on it."

But I didn't need to sleep on this decision, or at least I didn't think so. For months, I'd been feeling barbs of doubt and uncertainty—at first, only during certain moments. Eventually, I couldn't stop thinking about them: These questions that wouldn't leave me alone. The decision materializing before me that I knew could blow up my life. The decision I needed to make, because not knowing what to do, and floundering in that feeling, felt much worse than just doing something.

By that June night, my body was sending me a clear signal: Enough uncertainty. Enough doubt. Answer the question and do what you have to do.

That morning, as my husband rolled out of bed to make coffee, I slumped over to the couch and told him, tearily, that we needed to talk. As he sat down next to me with his steaming mug, a concerned look on his face, I told him that I'd been asking myself a question in the months since we'd gotten married, and I finally knew the answer. The question: Should we get a divorce?

This is not a book about marriage. It's a book about uncertainty. It is the book I would have liked to read when I was poised to detonate my relationship or at various other points in my life when I wrestled with huge existential questions and had no answers.

Often, during those moments, I would pick up books or listen to podcasts, seeking refuge in the latest self-help, pop-psychology, or New Age spiritual book to feel less alone or to find answers that might help me through a rough patch. Invariably, predictably, within those pages, I would encounter this phrase: *Embrace uncertainty*.

Typically, when this nostrum pops up in books, TED Talks, or magazine articles, it is meant to be a solution, a call to action for those of us who do not feel like nuzzling up to the uncertainty in our lives and giving it a big bear hug. If we could just embrace uncertainty, these advice gurus assure us—if we could let it settle in, bake it cookies, and invite it to stay a while—then we'd eventually banish our psychological demons.

But these gurus are missing the point. If we could embrace uncertainty, in all aspects of our lives, we'd also be a different species. We humans are programmed to try to dispel doubt and uncertainty, to seek the comforting bedrock of certitude.

This book is not about banishing uncertainty; it's about finding a different and healthier relationship with it, whether or not you want to embrace it. This book is also about the unexpected and often painful consequences we encounter when we try to rid ourselves of doubt by seeking fast, easy answers to some of our biggest questions.

The idea for this book emerged from my own humbling, messy journey to change my relationship with uncertainty during a tumultuous period of my life. It is a journey you'll witness throughout these pages, and one that continues today.

Maybe the phrase "embrace uncertainty" makes you, too, cringe, and yet you're struggling with big unanswered questions in your own life. Maybe you're asking yourself: Did I choose the right life partner? Who do I want to be? Will I ever live without pain? How am I going to survive the loss of a loved one? What is my purpose?

Welcome. You've come to the right place. We're in this together.

And there are countless others in this with you, too, many of whom will appear in the pages of this book. Some of them are public figures. You may not have heard of some others discussed here, but their stories will prove instructive: a wedding photographer who falls out of love with her fiancé; a woman paralyzed in a car accident rediscovering her identity; a professor with a mysterious illness who devotes her life to studying how to live better in uncertainty; an undocumented immigrant breaking free of questions that had been holding him back; a poet whose words more than a century ago give us a blueprint for living with our most challenging questions; and many, many others. For me, their stories are like fireflies, flickering during moments when I start to wonder whether anyone out there has ever felt what I'm feeling—whether anyone else understands. I hope that, after you read about the experiences of all these people, you'll know that they do understand what you're feeling. And so do I.

Unraveling Uncertainty

We use the term "uncertainty" all the time to talk about distinct life experiences—everything from waiting for a test result to wondering whether we've chosen the right partner. As a result, there are many ways to define the term. One early attempt by an economist in 1921 describes it as "the lack of knowledge about the probabilities of the future state of events."[1] Though undoubtedly accurate, this definition doesn't capture much about the *experience* of being uncertain. Perhaps that's why the definition that resonates most with me is that of psychologists Raanan Lipshitz and Orna Strauss in their 1997 paper about how decision-makers conceptualize and cope with uncertainty. In their words, uncertainty is "a sense of doubt that blocks or delays action."[2] Uncertainty, in other words, can freeze us in our tracks, leaving us to languish or stagnate. It can also impair our ability to make rational decisions. When we're faced with time pressure and uncertainty,[3] research suggests we tend to *exploit* options that are familiar to us rather than *exploring* new choices, even though the best choice is generally to alternate between exploiting what's familiar and exploring what's new.[4]

We relate to uncertainty the way we relate to ambiguity: not very well. Most humans, in fact, are programmed with ambiguity bias.[5] According to behavioral science research, most of us will do almost anything to avoid uncertainty, even if it means accepting greater risk or larger losses. Psychologists have named our tendency to want to resolve uncertainty "the need for closure"; this need applies to everyone, and it can change across situations.[6]

The need for closure can also change over time. Our intolerance for uncertainty appears to be growing.[7] One group of researchers has shown through a meta-analysis—a study of fifty-two different studies—that this trend is correlated to increased internet and mobile phone usage. Even though it might seem that having a device that can offer us answers at any moment would reduce anxiety, these researchers found that it paradoxically appears to have done the opposite for many people over the long term. We use our phones to seek reassurance and to feel safe when we're faced with an unpredictable experience: For instance, if we don't know which restaurant to go to in a new city, we can review Yelp ratings. In a moment of interpersonal awkwardness or while waiting in line, we can scroll through social media. As these behaviors harden into habit—*experience uncertainty, reach for phone*—we start to reduce our "spontaneous, everyday exposures to uncertainty,"[8] the kind that could make us more resilient to, and tolerant of, the unknown. Unsurprisingly, anxiety is also on the rise,[9] along with the myriad physical ailments that come along with it.[10]

In some contexts, of course, our desire to reduce uncertainty wanes or disappears altogether. We may even hunger for uncertainty, especially if we know it will eventually be resolved. Think, for instance, of sports or other competitive events. We'd find watching less exciting if we knew the outcome ahead of time. Even in our professional lives, a dollop of uncertainty—for instance, not knowing whether a new experiment will work—can be invigorating. And judging by the true crime obsession that pervades Netflix, books, and podcasts, many of us love a good unsolved mystery, unless that mystery is playing out in the context of our own lives. In that case, we often want to turn to the final chapter, or click to the final episode, to see how everything comes together at the end.

The Consequences of Pursuing Certainty

Our instincts to reduce uncertainty don't always serve us well. Instead, the obsession with pursuing what's known at all costs can make us vulnerable to industries designed to serve up fast, easy answers.

We pay big money to astrologists, life coaches, spiritual gurus, mediums, and other consultants to give us answers to Big Life Questions: where we should work, whom we should love, how we should spend our time. Research suggests that belief in astrology as a science is on the rise.[11] Burned out from online dating, more people are turning to matchmakers,[12] fueling demand for their services and inspiring new businesses.[13] And the coaching industry is growing fast, too—the number of coaching practitioners in 2019 was up 33 percent from 2015 estimates, according to a study from the International Coaching Federation.[14]

The need to seek answers is drilled into us by teachers and other caregivers from childhood. This encouragement isn't necessarily bad, particularly when the source is dispensing sound, evidence-based advice or when people help you uncover answers within yourself (as great coaches and therapists are trained to do). The problem is sometimes the people we turn to don't have our best interests at heart. They know how to present just the right mix of sagacity and snake oil to hook even the most clear-eyed consumer. Our human desire for quick answers leads us to seek them in the wrong place, at the wrong time, from the wrong people.

But what are we supposed to do *instead of* seeking comfort in answers? This was a question I began to confront in my early thirties, at a point during the COVID-19 pandemic when the global mood was taking a more optimistic turn. In the US, the Food and Drug Administration had approved a vaccine, and people were tentatively booking flights or thinking about venturing out to play a game of trivia in a boisterous bar or to attend a concert indoors.

I, on the other hand, braced for impending doom. I couldn't shake a persistent and pernicious sense that I had just made two ruinous life decisions.

The first was to leave my job without having another one lined up. The second was to marry my boyfriend, now my husband.

These decisions, though exciting at the time, resolved nothing and only raised more questions: What did I want to do with my life? Would I ever find another job? Did I marry the right person? If I did, why was I so miserable?

And I *loathed* those questions—a strange experience for someone who claimed that inquiry was a scholarly passion.

My intellectual relationship with questions was easy. As an applied behavioral scientist and journalist, I'd spent years researching and writing about the power of questions in our personal and professional lives.[15] Questions, as an abstract subject of my work, were a convenient distraction, tucked inside a drawer of my mind I could open or slam shut as I pleased.

This was not the case for the queries that surfaced in my dreams, that nudged the dial of my heart rate, that were threaded through my every thought. These questions were my tormentors. I wanted answers—and I wanted them now. *How can I find a meaningful career? What's my purpose? Should I get a divorce?*

On a walk around my neighborhood in downtown Oakland, California, one evening, peering at the crepuscular pink sky and stuck in a raincloud of rumination, I heard a podcast extolling the wisdom of a poet named Rainer Maria Rilke and his book *Letters to a Young Poet*.[16] It sounded like a salve for anyone who was struggling with big questions.

The Third Letter

Let's travel back in time for a moment, to 1903. Rainer Maria Rilke, an Austrian poet, was broke. Even worse, he worried that he had run out of ideas. At twenty-seven, he had already written more than a dozen books, though he had little traditional "success" to show for them. Penniless in Paris, the poet received a letter in February 1903.[17] It sparked a correspondence that would later influence the lives of thousands around the world,[18] including such varied figures as Marilyn Monroe, Konrad Zuse (the inventor of the programmable computer), Dustin Hoffman, and Lady Gaga.

The writer of this crucial letter, Franz Kappus, was a nineteen-year-old enrolled in a military academy, an aspiring poet, and a fanboy of sorts. Kappus was asking the older poet to help him navigate the mounting uncertainty of life.[19] In his third missive, the religiously conservative Kappus interrogated Rilke about whether sexual love was a sin, asking forbearance if he had rambled. He concluded, "I only wanted to share just one of the many questions leading their lonely existence in my soul."[20]

What's unusual about Rilke's response, and what has made it so enduring all these years later, is that he did not suggest that he had answers, nor that Kappus should seek them out. Instead, Rilke wrote, "I want to ask you, dear sir, as best I can, to have patience about everything that is still unresolved in your heart; try to love *the questions themselves*, like locked rooms, like books written in a truly foreign language. Don't look for the answers now: they cannot be given to you yet because you cannot yet live them, and what matters is to live everything. For now, *live the questions*. If you do, then maybe, gradually, without your realizing it, some far-off day you will live your way into the answer."[21]

Rilke was not explicit about what he meant by living the questions, but I've come to define his response as a strategy to break through the paralysis of uncertainty. It's a method of truth-seeking that is less interested in a final answer and more in the search for it, in what we can find along the way. Characterized by patience and openness to new ideas, those who live the questions crave the experience of continuous discovery, rather than a specific outcome. In his letter, Rilke converted what we might think of as wallowing—the static discomfort of being unsure—into authoritative action. He imbued the act of patiently holding a question with a sense of momentum, vibrancy, and vitality.

These lines—"try to love the questions themselves," "live the questions"—felt familiar and foreign to me at the same time. They comforted and challenged me. They piqued my curiosity because *loving* the questions felt decidedly different from *embracing uncertainty*. The idea of embracing uncertainty not only felt impossible to me, as I considered the potential dissolution of my marriage and my future career; it also seemed out of reach for so many other people dealing with their own uncertain experiences. Consider someone waiting to hear whether their cancer has come back, someone uncertain about the results of genetic testing on their unborn baby, someone not sure they'll walk again after an accident. "Embrace uncertainty" is the mother of all platitudes.

Rilke's emphasis on love, however, felt more realistic, more attainable. Love, after all, is hard work. It's often painful. In some situations, the object of your desire may act in ways that feel insensitive, aloof, or uncaring. They may seem unrecognizable from the person you fell for. In those moments, part of loving them is to be curious about *why* they're acting that way and to listen to their response even when you'd rather throw up your hands.

Embedded in this method of living is a recognition that it isn't always easy, that you may not want to love your questions, though you still must live with and commit to them. By continually returning to your biggest questions about relationships, meaning, purpose, and identity, you keep an important set of conversations alive: conversations with yourself, with those around you, and with the world.

But let's come back to earth for a moment. This was a nice idea, in theory. Still, I wondered: Is it really possible to love the questions of our lives—particularly the most painful and significant ones? What does a life of living the questions look and feel like? And what can we all learn from the people and communities following this path?

Throughout these pages, you'll meet many others who have taken a radically different approach to how they commune with uncertainty. And here, I'll offer an important caveat. Many authors have defined frameworks and offered up methods to guide readers toward decisions with resolve. But in this book I suggest that we should hold answers at bay—at least for the most significant questions in our lives, and at least for a little while. In the spirit of Rilke, we can learn to avoid the addictive appeal of fast, easy answers. Loving the questions is a practice that's seldom pursued, challenging to sustain, and utterly transformational.

Defining "The Questions"

There's probably a question nagging at you right now: What questions are we talking about here? The field of possible questions to love is vast. How do we distinguish among questions as varied as "Should I date this person?," "How can I be happier?," "Where should I live?," and "Who am I becoming?"

Throughout this book, I'll encourage us to think about all of the questions in our life like parts of a fruit tree.

PEACHES: Some questions are like fruit that ripens and can eventually be picked. These questions fall into the *answerable* category. There are *short-term answerable* questions; think about these as fruit that ripens relatively quickly, like a peach. These are the questions that, over a short period of time, get resolved—questions like "Will I get the job?"

PAWPAWS: It takes a pawpaw tree five to seven years to grow its tangy, yellowish-green fruit.[22] These, then, represent *long-term answerable* questions. Pawpaw questions eventually bear the fruit of an answer, but the time horizons are longer. For instance, you might question whether you should stay in a relationship, whether a new fertility treatment will help you get pregnant, or whether you'll end up liking a career change.

HEARTWOOD: Other questions are more like the inner wood of the tree.[23] The purpose of heartwood is to help the tree with balance, stability, and security. Heartwood questions are the ones that stay with you throughout your life. Though they're fundamentally unanswerable in a permanent way, they are your lifelong companions, helping you return to a sense of equilibrium. Think about questions like "Who am I?" and "How do I live a life of meaning and purpose?"

DEAD LEAVES: Finally, there are questions that do not ripen into answers and that we do not want to carry with us. These are the questions we eventually need to release. They no longer nourish us and are often questions about past decisions that drive patterns of rumination and suffering. These might be questions that emerge from regrets, about why relationships ended, about wanting closure that won't come: "Why didn't I just take that job?" "Why did we break up?" "What if I had just done it all differently?" If a question isn't helping you grow and move forward, it's time to let it go.

What to Expect Next

This book unfolds in three parts, each designed to make concrete a piece of Rilke's philosophical wisdom and to help us learn as much from the poet's story as we do from his words.

Part One, "Don't Look for the Answers Now: How the Search for Easy Answers Fails Us," looks at why we crave certainty, how our current culture has exacerbated this desire, and how the tendency debilitates our minds, our bodies, and our societies.

In Part Two, "Love the Questions Themselves: The Rewards of Committing to Curiosity," we learn about the many benefits of living in questions, and we trace the stories of people whose lives were transformed when they changed their relationship with uncertainty.

Part Three, "Live the Questions: Starting a Questions Practice," offers ways to begin living in questions yourself: evidence-backed tools to develop your own questions practice and a questions map to help you navigate your own uncertainty. You'll find inspiration for different questions to ask yourself and ways to approach what might seem like a mysterious pursuit: *live the questions now, and love the questions themselves.*

The pursuit seemed mysterious to *me*, at least. As I faced my own questions about my relationship, my job, my *life*, I felt a desire for two things: Yes, I wanted answers. Increasingly, though, I was also curious about whether I could live up to Rilke's ideal, learning to cherish the questions that were keeping me stuck. It seemed to me that if I *could* manage to love them—or even just hate them a little less—that maybe I'd feel better, less anxious and afraid. Maybe I'd feel less hungry for answers. And maybe the practice would unlock some deeper wisdom about how to live a more fulfilling, purposeful life.

Or . . . maybe it would teach me something else entirely. I'd tried almost everything else: What did I have to lose? It would, I decided, be a kind of philosophical experiment.

I hope my experiment will show you new ways of relating to uncertainty, using it to propel yourself toward self-growth and discovery, instead of staying painfully stuck. With deeper awareness of the questions you carry, I hope you'll also learn to distinguish between the ones you wish to keep permanently in your pocket and the ones that, eventually, you decide to let go.

Perhaps your questions will morph into new shapes and encourage the sort of self-discovery that's only possible when you feel most alive—that is, when you're open to all the possibilities of what could happen next.

PART ONE

DON'T LOOK FOR THE ANSWERS NOW ➤

HOW THE SEARCH FOR EASY ANSWERS FAILS US

1

The Charlatans of Certainty

THE BIG IDEA

Why are we so often seduced by fast, easy "answers" to our most significant questions? Though the reasons are complex and individual, this seduction is primarily driven by "gurus" who don't have our best interests at heart.

Rainer Maria Rilke is twenty-one and has no idea what to do with his life or even how to live it. What he knows is that he has a crush on a woman he's never met—someone he fell for after reading her writing. For weeks, he sends her a series of unsigned poems. Then he asks his friend Jakob Wassermann to introduce him to the woman. So Wassermann arranges for the two to meet at his apartment in Munich for tea.

The woman arrives, wearing a loose, cottony dress, with her wavy brown hair in a messy top bun. She tells stories unconventionally, with events out of order. Rilke is mesmerized. The woman is Lou Andreas-Salomé, a Russian feminist, writer, philosopher, and psychotherapist. She has already declined marriage proposals from Frederich Nietzsche and the philosopher Paul Rée.

First Rilke and Andreas-Salomé become lovers—and then something else. Rilke is obsessed with Andreas-Salomé; she's more skeptical. But she believes Rilke can be molded, groomed, improved. And Rilke is desperate to be shaped—for someone to tell him how to be, what to do, how to live.[1]

"I am still soft, I can be like wax in your hands," he writes in an autobiographical story after meeting Andreas-Salomé. *"Take me, give me a form, finish me."*[2]

"You've been having some problems with your boyfriend, haven't you?" the astrologer asked me over the phone—a raspy, unfamiliar voice in my earbuds.

Well. How to answer such a question?

I paused, remembering our fight the previous day. The unresolved tension polluted our living room, where I was taking this call now.

It wasn't that the fighting was new; our relationship had been challenging for years. I'd almost broken up with my now-husband, then-boyfriend, twice. We were different in significant ways. He'd grown up in a lower-middle-class family in Central California; I'd grown up upper middle class in the suburbs of Chicago. He was part Lebanese, and his parents raised him Catholic. I was Jewish. We'd both experienced childhood traumas, making us more sensitive and reactive in certain contexts. We often behaved in ways that triggered each other, creating what could feel like a never-ending cycle of fighting and hurt.

It hadn't always been that way. We'd met on an online dating app back in the spring of 2016. What I noticed first about his dating profile—which the algorithms served up during my first week in San Francisco, where I'd moved from Washington, DC—was that "A." was an English teacher. This seemed like a refreshingly different occupation in San Francisco, the tech bro capital. On our first date, we met at a bar in the Mission District known for its free peanuts. Shells littered the sticky floor. The bartender overfilled the drinks, and I accidentally spilled some of my dirty martini on this date of mine. I was reminded of the time I'd dropped a slice of pizza—cheese side down—onto the sleek silk dress pants of a DC date. ("Jesus, are you kidding me?" that guy, a lawyer, had hissed, dabbing his grease-stained crotch while I looked on in horror.) But A. just laughed. At the end of the evening, he asked if it would be okay to kiss me. It was, and he did—politely and passionately.

On our second date, we spent nine hours walking through the city, looping our way from the eucalyptus-lined paths of Lands End through the foggy Richmond District. On the fourth date, we went to a Prince-themed dance party, and he looked at me with so much tenderness as we swayed to "Purple Rain" that I thought I would explode. Sure, part of this was lust—the honeymoon phase of a new relationship. But something else drew me to him.

A., unlike other men I'd dated, asked me questions and listened closely to my answers. Also unlike other men I'd dated, he was curious not just about the polished, finely tuned version of me I presented to him, which I was accustomed to serving up. He wanted to know about who I was underneath all of that—a question I didn't yet have an answer to. This subterranean part was a version of me that was raw, ragged, unpredictable, angry. This version felt trapped by expectations about the way I was supposed to behave to keep my reputation and relationships intact.

A. didn't have the same kinds of expectations. In a way that was often uncanny, he could tell when I was saying something just to placate him or doing something because I thought I was *supposed* to. *This* was what upset him most. He expected me to behave not flawlessly, but honestly—even when it meant revealing thoughts and behaviors that felt ugly and unknown to me. Rather than turning away from these parts, he leaned in, wanting to know more and to understand dimensions of me I didn't yet understand myself.

For a few months, I was convinced he was the man I was supposed to marry. This, after all, *felt different*. And wasn't it supposed to feel different when you *just knew*?

But my doubts started creeping in at a close friend's wedding, the first one we attended together a few months after we began dating. At the intimate rehearsal dinner, most people were having a drink or two, if that. A., on the other hand, had decided to go not just a little harder than everyone else, but a lot. He had close to eight drinks before I pulled him aside at the bar. "What are you doing?" I hissed. He was sloppy, almost falling over. Furious, I dragged A. back to our hotel, wondering why he'd behaved this way when he was meeting some of my best friends for the first time. He was thirty years old, confident, and brilliant. What was wrong with him? Where was the empathetic, funny, and measured guy I couldn't wait to spend time with?

As I introduced him to other friends, the pattern repeated itself. He would drink too much or he would start a conflict—challenging someone's perspective on current events, for instance. Sometimes he would do both. I began to dread taking him to parties, where he would transform into someone I didn't recognize. It was years into our relationship when *he* finally understood, and explained to me, the reason for his behavior: insecurity. He felt intimidated by these people who he knew had come from wealthier backgrounds, who seemed to be more successful than him. Instead of grappling with those feelings of inferiority, he lashed out.

A couple of months after the wedding fiasco, he was laid off from his part-time job as a teacher, a role he'd chosen so he would have more time to write fiction. He wasn't sure what he was going to do now. In my early twenties, I wouldn't have sweated this uncertainty. But now I was evaluating him as a potential life partner. I wanted a mate who would be stable, who could help pull some weight financially, who could get along with my friends. Up until then, he'd prioritized time for his creative pursuits over money. *Did our priorities match up?* I wondered. *If not, was this doomed?*

And there was another emergent quality of our relationship that felt hard to articulate. Sometimes, his confrontational nature triggered deep insecurities about my intelligence and competence that, despite years of therapy, I had yet to successfully address. Like me, he enjoyed intellectual conversations. Unlike me—and unlike anyone else I'd ever dated—he was argumentative and unafraid to challenge my ideas and perspectives. Even though *he* never questioned my intelligence, I worried that he didn't think I was smart or interesting enough, and I subsequently became nervous when we explored unfamiliar topics of conversation.

His tendency toward confrontation came with a stark trade-off. On the positive side, he was encouraging me to grow in a way I never had with another mate, by identifying how my people-pleasing traits made it hard for me to be authentic, even with people I thought of as my best friends. On the negative side, his assertions could be unsettling and upsetting. We were regularly having uncomfortable conversations and even fights. I had never fought, really, with other boyfriends—or with anyone else in my life. The relationship raised deep questions about what I was drawn to and what I was afraid of.

For years, our relationship was a seesaw of certainty and doubt. Even though we'd had periods of fighting before, this stretch—just before I spoke to the astrologer—felt different. We'd moved in together a few months earlier. Living together had started out promising, the new space offering a kind of relationship renaissance as we picked out bedspreads and delighted in joint grocery shopping. But now, after the honeymoon phase, I wondered whether it had all been a huge mistake. I found a therapist, whom I regularly told, "I don't know what to do," as if she could, or would, give me the secret solution. I divulged my struggle to a few friends, hoping that someone, anyone, could relate. Most friends had chosen partners who were more similar to them and couldn't quite understand why I would stay with someone who seemed to stress me out. After these conversations, I often felt even more alone.

Though I didn't admit it to myself at the time, the act of making an appointment with the astrologer was a Hail Mary. For years, I'd told myself that I didn't believe in astrology, psychic readings, or tarot cards. But my actions in seeking out these practitioners told a different story, one of a person torn between what she *wanted* to believe and what she knew to be true. I *wanted* to believe that these practices revealed deeper truths and could help me. And yet, as a science-minded person, I knew there was scant evidence for such beliefs. In this case, it was easy to continue to delude myself and downplay my own agency. The reading was a gift from a well-meaning family member, so I wasn't even paying for it. I told myself that it wouldn't amount to anything more than some light entertainment.

The reading, however, was not very entertaining. As I spoke to the astrologer, I realized I was holding my breath. "We're going through a rough patch," I told her, not wanting to give too much away.

She was quiet for a moment. "Listen," she began slowly. "I usually don't like to tell clients something this direct. But from what I can see in your charts, he isn't the one for you." Another long pause. "It's not going to work out between you two."

Staring out the window of our stuffy Oakland apartment, I felt a series of unsettling questions snake through my stomach and chest: *Is this an incontrovertible sign that we should break up or a load of BS? What the hell do I do now? If we do break up, will I ever find someone again? How can I hurt someone I still love? Do I still love him?*

Why was I so tempted to believe this person—the astrologer—whom I'd never met before, who was using a practice I didn't even believe actually worked?

And why, *years* after this reading, was this astrologer's voice the one playing in my head when I was considering whether to divorce my husband?

The Temptation of Fast, Easy Answers

"Fast, easy answers" may mean something different to everyone. I'll define them as answers that come not from a sometimes lengthy—and challenging—process of investigation, but from a short engagement, often with an outsider who claims that they know you and what you want better than you know yourself. These outsiders often claim they've discovered the secrets of life, how to simplify the complex, and how to erase suffering from our existence. Significantly, they may even discourage people from further exploration. It's as if these outsiders are producing answers for the masses, dispensed from a vending machine and designed for short-term pleasure and relief. One 2007 book title says it all: *Happiness Now! Timeless Wisdom for Feeling Good FAST*.[3]

Just as it's okay to indulge in a bag of potato chips occasionally, all mass-produced sources of support or guidance aren't automatically harmful. But if that's *all* you're consuming—if somehow you've been tricked into thinking that your potato chips are, in fact, as nutritious as broccoli—you'll end up malnourished.

In her book about the self-help industry, writer Jessica Lamb-Shapiro distinguishes between empowering and exploitative versions of self-help using the analogy of open and closed systems.[4] Closed systems are ones you can't escape. In these systems, the problem is always *you*, not the answer that's being proffered: *This thing works, and if it doesn't work for you, you're not doing it right.* "That always seemed sinister to me," Lamb-Shapiro told me, almost ten years after she wrote the book.[5] These systems exude defensiveness and may even discourage the asking of questions.

Lamb-Shapiro writes about attending *Chicken Soup for the Soul* author Mark Victor Hansen's MEGA Book Marketing University, where participants were ostensibly taught how to write a bestselling self-help book.[6] Along the way, Hansen told the audience, "Don't be a tire-kicker. If you're a tire-kicker, you're not going to get anything out of this." This admonishment, Lamb-Shapiro points out, was "very different from an actual university, where most students are encouraged to turn their critical faculties on both themselves and what they are taught. Further, in an actual university one eventually graduates; whereas here it seemed that the potential for attending seminars and buying inspirational tapes was endless."[7]

Open systems, on the other hand, exist when people offer advice and sets of tools with the encouragement to take what works and discard what doesn't. The message of these purveyors is "This may not work for everyone" or "This may not be your answer." Think about a great therapist or coach who shares a framework but ultimately helps to guide you to your own answer—or the leader of a workshop I recently attended, who told us on the first evening, "You might not agree with what I'm sharing, but I invite you to reflect on it."[8]

Indeed, journalist and author Claire Hoffman found that the

difference between being told what an idea should mean for you versus being invited to explore it for yourself is one of the key distinctions between an open self-help system and a closed one. Hoffman has some personal experience navigating the world of exploitative self-help gurus. She grew up in Utopia Park, the cultlike community designed around Transcendental Meditation and its founder, Maharishi Mahesh Yogi. In her memoir, she recounts the revelatory experience of discovering agnosticism as a teenager. "The idea that you could define yourself by the decision not to have an answer was so new to me," she writes.[9]

Now, if a friend who has recently returned from a self-help workshop or a yoga retreat claims they need a new language to talk about the experience, Hoffman's "You Are Joining a Cult" buzzers go off. "It's a vocabulary thing," she told me.[10] "When people use words that aren't words, or they're using a word over and over," Hoffman sees it as a sign that they may be on the path to joining a community that may not have their best interests at heart.

Linguist Amanda Montell came to the same conclusion. "From the crafty redefinition of existing words (and the invention of new ones) to powerful euphemisms, secret codes, renamings, buzzwords, chants and mantras, 'speaking in tongues,' forced silence, even hashtags, language is the key means by which all degrees of cult-like influence occur," she writes in her book *Cultish*.[11]

Today, cultlike groups "feel comforting in part because they help alleviate the anxious mayhem of living in a world that presents almost too many possibilities for who to be," she writes, recalling a scene from the Emmy-winning series *Fleabag*, in which Phoebe Waller-Bridge's character tells her priest, "I want someone to tell me what to wear every morning. I want someone to tell me what to eat. . . . I want someone to tell me what to believe in, who to vote for, who to love, and how to tell them. I just think I want someone to tell me how to live my life."[12]

Seekers of self-help are often united in contradictory desires: They are both seeking to reclaim agency over their lives and hungering to be told what to do. Perhaps you've felt that tension in your own life: You want to figure out the next step in your career, but "What should

I do with my life?" is such an overwhelming question that you also secretly hope someone will just *tell* you. Or you may hope that if you just take enough personality tests, the answer will become clear. I've been there many times. The problem is that we try to escape that uncomfortable feeling of not knowing by potentially making ourselves *more* vulnerable to those who don't have our best interests at heart. In other words, we're searching for answers in the wrong places.

It was perhaps no surprise, then, that a few hours into the Mark Victor Hansen seminar, Lamb-Shapiro, the self-help memoirist, started to realize her own vulnerability to "the pleasures of the imperative. I had been told what to do, wear, and eat for the last two hours, and it brought me a kind of comfort. It was no accident, I thought, that the last time I had this little responsibility for my actions I had braces on my teeth, a bona fide mullet, and someone to cook me dinner every night."[13]

Lamb-Shapiro was in her early thirties when she wrote the book. Now, in her mid-forties, she says she has a very different opinion of self-help. "I have a lot more sympathy for the existential terror that comes with confronting the fact that we can't control things," she told me. "The uncertainty in the world has gotten worse. It's not that I think self-help books are better, but I have so much more empathy for why they are wanted and needed and the emotion that goes into wanting to control things during a chaotic period."[14]

This was a feeling I could relate to. Since I was a kid, that existential terror is one I've tried to keep at bay. I've been drawn to people and disciplines that supply fast, easy answers—from the evidence-based kind to pseudoscience. Coaches, therapists, nutritionists, psychics, astrologers, tarot card readers—I've tried them all, nurturing an insatiable desire for answers often against my own better judgment. First in line to get my tarot cards read or my fortune told at parties, I always told myself it was *just for entertainment* (as I did with the astrologer) even though I knew I was hoping for more. After a particularly hellacious breakup in my mid-twenties, I knocked on the door of a DC psychic in Dupont Circle asking if I would ever find another mate.

At dinner with an astrology-obsessed friend, I found myself reading my chart in her astrology app and feeling a sense of comfort. I spent a week in Costa Rica at a jungle yoga retreat, trying to avoid the ever-present tarantulas and taking copious notes in the margins of a book that promised a blueprint for happiness, mapping out how I might jettison my negativity.

As often as I've been skeptical of gurus and influencers, I've also wondered, *But what if they really could tell me a better way to live?* I've scoffed at friends and family members who seemed to have bought into it all. Meanwhile, though, I couldn't help noticing how much I *wanted* to believe that my life's mysteries could be unlocked through one person's intuition, or a deck of cards, or a weekend of pseudo-therapeutic exercises—that someone else, or something else, might hold the key to the inscrutable, subterranean dimensions of my consciousness.

Why did I, and so many others, keep getting drawn in?

The Science of Seduction

When you think about seduction, what comes to mind? The common understanding of seduction is that it's something one person does to another, generally for the purpose of romance or sex. But we can also be seduced by ideas and beliefs. The word "seduce" has its roots in the fifteenth century and comes from the Latin *seducere*, which means "to lead away."[15] Fast, easy answers can trick us into thinking we've reached our destination, when in reality we've wandered off the path.

In 1936, the German psychologist Kurt Lewin helped us understand why we might wander off that path, even if we think we know better. Lewin's discovery was a revelatory insight in the "nature versus nurture" debate about human behavior.[16] It's an idea that remains one of the most enduring lessons of the social sciences.

Even when people grow up in similar environments, they can have very different psychological profiles, Lewin wrote in his book *Principles of Topological Psychology*. According to him, human behavior results from a complex and ever-changing interplay between who we are, where we are, and where we came from. Indeed, when it comes to understanding why any of us might be seduced by fast answers, we need to explore a set of interrelated influences: the biology we share as humans; our unique personality and upbringing; how secure or threatened we might feel; and who, exactly, is providing the answers. We can divide these influences into three big categories: who we are, how we grew up, and who's answering. Unpacking each one helps us better understand *why* we get seduced by fast, easy answers, so we can be more aware of these influences in our own lives.

Who We Are

Let's begin with the basic findings from contemporary evolutionary and social psychology and neuroscience: Multiple studies suggest that we prefer certain pain or loss to uncertainty and that we tend to avoid ambiguity at all costs.[17] We humans are *wired* to avoid or reduce uncertainty and to minimize surprise by changing the way we respond to our environment.

At any given moment, our brains are trying to use as little energy as possible—a smart adaptation for when food was more scarce and we had to curtail exertion.[18] Uncertainty, because it forces us to use more metabolic energy, spurs us to reduce ambiguity quickly to conserve future energy. This is known as the selfish brain theory. This metabolic readjustment can feel like stress or anxiety.

Psychologist Arie Kruglanski named the psychological part of this experience the "need for cognitive closure."[19] After decades of research, he along with researcher Donna Webster published a scale in 1993 to help sort out which individuals have a higher need for closure—in other words, a stronger desire to seek and then cling to answers. Anyone can take the scale assessment. Be prepared for prompts like these: "I think that having clear rules and order at work is essential to success" and "Even after I've made my mind up about something, I am always eager to consider a different opinion."[20] The scale assesses two key dimensions of our orientation to closure. First, in *which direction* do we lean: Are we approaching or avoiding answers? And second, *what kinds of answers* are we seeking: Are we searching for a specific answer, or will any answer do?

At its core, "closure is the elimination of possibilities, of imagination, possible worlds, things that could be," Kruglanski told me.[21]

When we're seduced by fast, easy answers, we're likely experiencing what he calls the "need for nonspecific" or "specific closure": we're approaching the answers, and we want them now. Experiencing the need for cognitive closure often creates in us a sense of urgency about getting an answer, leading us to "seize" on an idea and inducing "a kind of mental impulsivity in which one leaps to conclusions on the basis of skimpy evidence," Kruglanski writes in his book *The Psychology of Closed Mindedness*.[22] Instead of exploring other possible answers or options, in this state of mind we may also "freeze" on the answer we found, closing ourselves off to other exploration.

All of us, depending on the situation, may find ourselves freezing and seizing.[23] Maybe you're comfortable with uncertainty at work, and you can tolerate a longer period of not knowing a specific answer, even as you move toward one. But when it comes to your personal life, you might be more likely to want to seize and freeze—or avoid getting an answer all together. This is perfectly normal. Scientists have also found that during moments when we feel our sense of control is threatened, we seek *compensatory control* by seeing patterns in randomness, clinging to conspiracy theories, defending sociopolitical institutions, or doubling down on religious beliefs.[24]

How much we seize and freeze, and when, is also a function of our personality—and specifically, our level of openness.[25] Openness describes how receptive people are to new ideas and experiences. And in some cases it can be influenced by our upbringing and our experiences of trauma. When I look back at how I grew up, I suspect certain childhood experiences contributed to me becoming the type of person who might be more fearful of uncertainty and more desiring of fast, easy answers.

How We Grew Up

When I was eight months old, everyone in my family got the flu.

My mother, thirty-six at the time, did not recover. She could not get out of bed, open her eyes, or even speak. To conserve energy, she whispered or wrote notes. Months went by. Eventually, doctors diagnosed her with myalgic encephalomyelitis, also known as chronic fatigue syndrome. For my mother, the main symptom was a paralyzing exhaustion that left her debilitated and unable to care for me. For others, the syndrome can also include impaired cognitive function, sleep issues, and inability to tolerate sensory inputs. It is diagnosed in women two to four times more often than in men.[26] Years later, a psychiatrist would suggest to her that her condition could have also been due to postpartum depression.

As I grew older and she worsened, family videos featured me, the only child, bouncing through the rooms of our Dallas home. I approached my mother's bed and tugged her arm, attempting to resurrect my playmate. "Get up! Get up!" I pleaded. She murmured some placating excuse and stayed supine. Eventually, dejected, I slinked away.

For nearly five years, my mother saw doctors, therapists, and homeopathic healers, spending months at a time away from our home while a parade of babysitters read me bedtime stories, brushed my unruly curls, and slathered peanut butter on my toast. My father worked as a national correspondent for the *Chicago Tribune* and spent most weeks traveling on assignment. It was because of his job that my parents moved from Chicago to Dallas, Texas—stranding my mother far from her network of family and friends.

My mother was also a journalist, but she had stopped working full-time to care for me after I was born. It turned out the person she needed to care for was herself. "I function at about 20 percent of my old healthy self," she wrote in an article about her condition in 1992, when I was four. "The illness remains a medical mystery, and doctors can give me no prognosis."[27]

It took her years to discover that the "cure" for her condition was more complicated than rest or medication. Part of her paralysis was rooted in decades of denying the emotions she felt, the questions she had. Her authentic voice was submerged below the sediment of cultural and family expectations. She'd grown up walled off from her emotional truth, never asking herself the kinds of questions that might have led her closer to it, never even knowing what questions to ask. In a conservative Jewish family struggling to raise her along with her special-needs brother, no one ever asked her, "How is having a special-needs brother affecting you? How are you doing?"

"I didn't see myself as counting as a subject worth exploring," she said. Her experience didn't matter, she thought, so why should she have bothered excavating it?

As an adult, those deeper currents of inquiry remained invisible to her. "I felt like I couldn't question the role I was supposed to play as a cheerful, ever-present mother and a supportive wife, even when I was unable to keep up that facade," she told me recently. "My not recovering was a symptom of a deep distress and a profound sense of powerlessness—a powerlessness to ask questions."

She got better. But it took five years. She eventually returned to full health by the time I was in kindergarten; I began to forget her illness had ever happened.

And I rarely talked about it, or brought it up, until a therapy session when I was thirty-two.

It was toward the end of the session, and my therapist seemed vaguely perplexed by an emotional reaction I'd had to some recent events. "I know you've said you had a happy childhood, but sometimes when you describe your experiences, they sound like things I hear from patients who had severe childhood traumas," she said.

Up until then, I'd never mentioned my mother's illness—the five-year gap in my care—to her. In fact, it was something I rarely talked about with *anyone*. I was pretty sure it hadn't affected me all that much. After all, I hardly remembered it. *I was fine.*

As I began to divulge these details to her, I could see a mix of shock and understanding flash across her face. "Oh, my god, Elizabeth," she said quietly. "This is huge."

I'm hardly alone in carrying around a challenging, and largely buried, childhood trauma. But since I maintained an outdated definition of what counted as trauma, I'd spent years dismissing my own. In his book *The Myth of Normal*, physician Gabor Maté argues that it's time to challenge our ill-defined notions of trauma and who has been traumatized.[28] It's a term, he says, that has been confused and diluted in colloquial contexts.

"The usual conception of trauma conjures up notions of catastrophic events: hurricanes, abuse, egregious neglect, and war," he writes. "This has the unintended and misleading effect of relegating trauma to the realm of the abnormal, the unusual, the exceptional. If there exists a class of people we call 'traumatized,' that must mean that most of us are not. Here we miss the mark by a wide margin. Trauma pervades our culture. . . . In fact, someone without the marks of trauma would be an outlier in our society. We are closer to the truth when we ask: Where do we each fit on the broad and surprisingly inclusive trauma spectrum?"[29]

Maté suggests that there are two broad categories of trauma—capital-*T* trauma, which we might think of as that *exceptional* trauma category—those who have survived war, natural disaster, and abuse. This is the trauma experienced when "things happen to vulnerable people that should not have happened," he explains. But then there is the lowercase-*t* category of trauma, which includes experiences like bullying or a lack of emotional support from caregivers. "Children, especially highly sensitive children, can be wounded in multiple ways: by bad things happening, yes, but also by good things not happening, such as their emotional needs for attunement not being met, or the experience of not being seen and accepted, even by loving parents."[30]

Trauma, Maté writes, is "not what happens to you, but what happens inside you . . . a psychic injury, lodged in our nervous system, mind and body, lasting long past the originating incident(s), triggerable at any moment."[31]

What unites all these forms of trauma is that they are forms of disconnection, of fracturing, he says. Trauma "is about a loss of connection—to ourselves, to our bodies, to our families, to others, and to the world around us" according to Peter Levine, the author of multiple books about trauma and the creator of a methodology for healing called Somatic Experiencing (SE).[32] When people experience a traumatic episode, the emotion can live in the body for years without being acknowledged and processed. The idea behind SE is to develop a deeper connection to bodily sensations and emotions so that trauma can be identified and released.

If we don't process our trauma, it can continue to warp our sense of safety and security as we get older, creating the illusion of threats where there are none and inducing a kind of psychological paralysis that limits our capacity for personal growth.[33] In this way, it has a profound influence on our tolerance of uncertainty and whether we see the future as threatening and hopeless, as full of potential, or as something in between—whether we desperately crave answers or are comfortable holding them at bay.

Who's Answering

Sophie was aching for answers when she attended her first meeting of the Olympus Group—a global personal development organization that offers adherents the promise of empowerment and transformation through its learning programs. Its website claims to cycle thousands of people through its courses each year, and it has a presence in cities around the world. Sophie had gone to the meeting in London because a friend from work encouraged her. The friend confessed that she was worried about Sophie, whose anxiety regularly tailed her into the office.

At the time, Sophie worked at a charity that mentored and tutored young people who struggled in school. Her friend was right: Sophie was depressed and unsettled. She'd recently moved to London from a city in northern England and didn't know many people apart from her work colleagues. On most weekends, she saw her long-distance boyfriend—a guy she describes as kind and caring but whom she could never quite trust because of her own insecurities. On Mondays, she'd trudge into work obsessively worried about the state of their relationship: Had the weekend been fun enough? Was the relationship the right one? Would he eventually dump her because of the long absences?

"For a person who thinks they aren't good enough, being in a long-distance relationship is hard," she reflects, now in her late thirties. Sophie had struggled for years with eating disorders, anxiety, and sometimes paralyzing self-doubt. The night she attended the introductory Olympus meeting with her friend, she was showered with love, encouragement, and a kind of manic energy. "It's like a fervor, she said. "It's so energized and so electric that it's almost like your nervous system can't help but be excited about it as well." Everyone was happy. Everything was amazing. She was unaccustomed to being surrounded by so many joyful people. "I was like, I want what these people are having," Sophie recalls.[34]

But more than that, what the leader of the session shared made an impression on Sophie, validating the feelings and experiences she'd been having and persuading her that these people might actually be able to heal her. The Olympus leaders explained that often we suffer because of the stories we tell about ourselves. Something happens to us, and we attempt to explain why it happened; then we end up clinging to that story and trying to find more evidence for it, even if it isn't true. So if your story is "I'm not good enough," as Sophie's was, you might interpret events across your life as supporting that story, even if another explanation is just as plausible.

Nothing about this approach is novel, Sophie later learned. But at the time, it was new to her and seemed to speak directly to her experience. She felt a sense of elation: Maybe, all this time, she'd simply been concocting the self-defeating idea that she wasn't good enough.

"If I had gone into the room and everyone was telling me I have to wear a purple hat for the rest of my life, I would've been like, I don't think this is for me," she said. "But they really touched onto the core things I had been dealing with through my life. I was like, this has to be a good thing." Through the group exercises, Sophie began to "imagine a life where I was free from doubting myself and feeling like I wasn't enough. It felt so clear that that was achievable by taking their programs." It felt clear because the leaders explicitly promised that taking the Olympus programs would allow participants to begin living an enriching life with more freedom and power than they'd ever known. The leaders also emphasized that "life was ticking away." As the minutes passed, you were left with less time to live the life you wanted to live.

If she'd experienced this much insight during just one evening session, Sophie thought, imagine what she could learn by attending the weekend retreat? So she did—twelve hours a day for three days. After the weekend was over, though, Olympus staff told her that her journey was just beginning. More answers awaited her—in the next program, for another 450 pounds. Sophie didn't have the money at the time, but the staff told her not to worry—that if she figured out how to make the investment in herself, the money would come. That, they said, was how the universe worked.

This type of magical thinking is a major red flag.

Sophie called her father, telling him about her experience and asking if he might lend her the money for the course. Her parents had always been supportive of her. But Sophie's dad was skeptical. The money, he told her, wasn't the problem. The issue was how quickly it was all happening. "Didn't you just complete the other program?" he asked. What was the rush? "Couldn't you take some time to think about this?"

Olympus staff had prepared Sophie to field these questions, which they said were bound to come from people in her life who wanted to stifle her development and keep her small. Finally, her father acquiesced. "I still remember him saying, 'If there's a problem, I'll come get you,'" Sophie recalls, choking up.

Sophie was hooked. But here, she told me something important:

Not everything she learned was harmful. That's the thing about places like Olympus. "There will be some things that actually work or make a difference for you." The wisdom can be mixed up with the nonsense, making it harder and harder to discern truth from lies. And "the more and more you go through it, the value is drained because you're giving more than you're getting."

After a year of attending every session she could, Sophie quit her nonprofit job to become a full-time paid staff member at Olympus. She began putting in eighty-hour weeks, doing what she describes as glorified telemarketing to get people to register for programs. She had to attend meetings every two hours to ensure she was meeting her numbers—lest she be publicly berated or worse threatened with losing her low-wage salary.

"I was becoming someone who was even more hypervigilant and anxious," she said. The more time she spent at Olympus, with her Olympus colleagues, the more she saw the world through the lens they gave her. And the more she found it hard to connect with anyone who *didn't* see the world that way. "It stunted my ability to have relationships with people who weren't involved in Olympus," she said. "I felt like I didn't have common ground with them anymore."

Her circle shrunk, along with any self-confidence she might have started to develop at the beginning. "It took me into a place of more self-doubt," she told me.

About two years after she joined, she started to get migraines. One day, as she waited on the London Bridge platform for her train into work, her world went black. Sophie passed out on the platform. The next thing she remembers was that a few train staff and paramedics were coming down to get her with a wheelchair. She could barely open her eyes and was having trouble speaking. At first they thought she'd had a seizure or a stroke. But at the hospital, a neurologist ruled out those possibilities, telling Sophie that what she had was a severe type of migraine that caused a partial paralysis on the right side of her body. For months, she didn't have full movement in her right arm and couldn't hold a fork to her mouth without shaking.

The neurologist began asking Sophie questions about her lifestyle and her work, trying to find clues about the cause. As she told him about her job, he was stunned by how much she'd been working. "He was like, 'Do you know you're working more hours than I was when I was a medical trainee?'" she recalls. This was a wake-up call. As she continued her recovery, she also quit her job at Olympus.

"So many people can be seduced into joining groups that promise the answers to all of life's most complex challenges," she said. "But the truth is that no group, religion, or person could possibly have all of the answers."

It can sure seem as if they do sometimes—and sometimes, that's what we want to believe. But as Sophie's story shows us, our tendency to be seduced is preventing us from finding our answers and is keeping us stuck.

The Seductive Power of People

"Modern day science makes clear that unpredictability has far reaching consequences for the lives we can envision and create for ourselves," writes the psychologist Scott Barry Kaufman in his book *Transcend*.[35] We're able to explore, grow, discover, and reach our full potential only if we can meet other needs first: safety, certainty, predictability, coherence, continuity, and trust in our environments.

That might sound like a lot of boxes to check before we're cleared for boundless self-discovery. But of all of these, Barry Kaufman emphasizes our experience of *coherence* as most connected to whether we feel safe enough to explore. For us to feel coherence, we need to think that our immediate environments make sense—that they are predictable and comprehensible.

Something that helps us do so is developing strong relationships. Our relationships play a powerful role not only in mediating the levels of threat and uncertainty we feel, but also in determining whether we'll believe an answer in the first place. Indeed, we're often seduced less by the answers themselves than by the people delivering them, as

Sophie discovered at Olympus. We are hungry to feel cared for, valued, and seen for who we are. People who satiate these needs gain our attention and trust—whether it's deserved or not.

When we decide to believe someone's answer, we're deciding, fundamentally, whether we trust them and whether the answer seems right. This is called epistemic vigilance, the process we go through to decide whether to believe a person, an information source, or both.[36] To make these calls, we often rely on multiple cues—many of which, according to the cognitive psychologist Tania Lombrozo, are imperfect at best and misleading at worst.[37] Confidence is one of the most powerful of these cues. Unfortunately, how confident someone is when they're expressing an answer does not tell us much about whether they're *right*. But their confidence can make us feel more comfortable and secure, which is, ultimately, what we're often seeking in the first place. And it's not just their confidence—we're also looking at a person's race, class, and language to determine trust.[38]

Under ideal circumstances, with plenty of time and cognitive bandwidth, we evaluate whether we can trust someone based on three main characteristics.[39] First, we assess their *competence*: Do they seem to be in control of the situation? Second, we evaluate how we perceive their *integrity*: Do they act the way they say they will? And, finally, we think about their *intention*: Do they seem benevolent?

More often, though, we make these judgments quickly, relying on imperfect heuristics, or rules of thumb.[40] Frequently, when we want a fast, easy answer, we may be feeling less than our best, buckling under the weight of a big question, and seeking relief and resolution yesterday. So we'll focus on how attractive the person is, how intelligent they seem, whether or not they seem aggressive or calm and welcoming. As with other behavioral biases, we may subconsciously rely on these heuristics because we feel a greater sense of threat or insecurity in our lives and a need for cognitive closure.

These rules of thumb don't always explain why we might get seduced, however, according to London School of Economics sociology professor Eileen Barker. She showed through her research into the

Unification Church (a cultlike group) that the people more likely to stay in the group were not those we would consider to be more vulnerable, such as people who had traumatic childhood experiences or who were less intelligent.[41] On the contrary, those who stayed were highly intelligent, optimistic, good-natured, and often oriented toward activism. They were people who already believed in the possibility of a better world.

"In this way, it's not desperation or mental illness that consistently suckers people into exploitative groups," writes the linguist Montell. "Instead, it's an overabundance of optimism. It's not untrue that cultish environments can appeal to individuals facing emotional turmoil . . . but the attraction is often more complex than ego or desperation, having more to do with a person's stake in the promises they were originally told."[42] In other words, how much do a leader's promises matter to you? How well do those promises apply to *your* life and *your* circumstances?

The Charlatans of Certainty

Leaders of cults, like the leaders at Olympus, can and often do tie their promises directly to the deepest and most meaningful questions of the people they wish to seduce. This is a key strategy of those I call the Charlatans of Certainty. You can think of the Charlatans of Certainty as people who are purposefully lying or manipulating others into believing that the Charlatans have all the answers, often swindling others in the process. The Charlatans of Certainty are not always bombastic, ostentatious gurus or nefarious psychics, making outrageous claims and bilking trusting followers of their money. Some Charlatans are far more subtle.

Take the proliferation of personality tests, each with its own promise of taking you deeper into your strengths, weaknesses, motivations, and identity. These tests can be helpful—especially if a test-taker feels as if a test expanded their self-awareness. But categorization of traits is a powerful tool that isn't always used responsibly, especially when those categories affect not only how you see yourself but also how you see others.

Jay Medenwaldt first discovered the Enneagram in 2017, while he was attending seminary and volunteering in a university psychology research lab.[43] The test likely dates back to 1954, though its origin story is a little hazy.[44] Some experts say it emerged as early as 2500 BCE in Middle Eastern oral traditions. Others claim the Russian philosopher George Gurdjieff learned of it while traveling in Turkey or Afghanistan. The most solid evidence places its emergence in the West in 1954, when Bolivian psychologist Oscar Ichazo started to weave ideas from the Enneagram into modern psychological theory, eventually teaching his system to psychiatrist Claudio Naranjo.[45] Naranjo then began to translate Enneagram insights into Western language and share them with students.

As of 2023, the hashtag #enneagram had 370.4 million views on TikTok.[46] "At its core, the Enneagram helps us to see ourselves at a deeper, more objective level," the Enneagram Institute® website claims.[47] In essence, it's a personality test that supposedly reveals our inner motivations by assigning test takers a number between 1 and 9 that lines up with one of nine personality descriptors—among them, the achiever, the challenger, and the investigator.*

When Medenwaldt first heard about the test, he was surprised and a little embarrassed that he hadn't been aware of it.[48] After all, he held an MA and BS in psychology and had served for nine years as a behavioral scientist in the air force, where he also taught psychology. He researched the test and tried to figure out why he'd never come across it. The answer: It isn't scientifically validated and wasn't in any of his textbooks about psychology or counseling. This was not on its own a reason to discard the test, so he dug into the scientific literature assessing the Enneagram's validity—whether it measures what it promises to measure. In doing his own research and evaluation of the

---

* In case you're wondering, I've often struck a tie between a 3 (the achiever) and a 9 (the peacemaker).

test, which he regularly presents on at conferences, he found that "the Enneagram is not a valid way to describe people."[49] This conclusion is reflected in a 2021 systematic Enneagram literature review, which found mixed evidence for the test's validity, while also finding that much of the Enneagram literature out there isn't published or peer reviewed.[50]

As Medenwaldt began to do his own analysis of Enneagram research and to discuss the test with more psychologists in his field, hearing their stories and experiences with it, he began to think the test could actually be harmful for some who use it. "People have privately told me that being the wrong type prevented them from getting a job, caused a breakup, and in one case, a divorce," he shared in a conference presentation on his work. "Others have experienced anxiety and shame because the Enneagram didn't work for them. Unfortunately, many people suffer in silence because they assume they're alone and fear social repercussions."[51]

One reason the test is potentially harmful, Medenwaldt argues, is that there's no way to discern the accuracy of its assessment. "It's supposed to reveal things about yourself that you are not aware of, but there is no way to tell if it is wrong," he writes. This could prove harmful in a couple of ways: first, if you accept it uncritically but it's wrong and, second, if you disagree with it but it's accurate. "There's no way to actually know!" he points out.[52]

The test is also an easy way to (inaccurately) stereotype people around you. Once you identify someone else as belonging to one of the numeric categories, you may believe you understand much more about their motivations and intentions than you actually do. And that's assuming the person's assessment is accurate in the first place. Indeed, we already tend to attribute someone else's behavior to their personality or character, rather than situational factors. This is known as the fundamental attribution error.[53] Consider a situation in which a colleague is late to an important work meeting. *He's so irresponsible*, you might think to yourself before you know the full context. In reality, your colleague was late because his kid's childcare fell through, and he had to scramble to find someone at the last minute.

"The best way to understand the mind of another person is just to ask them what they're thinking and feeling at the time," says Nick Epley,[54] University of Chicago psychology professor and author of *Mindwise: How We Understand What Others Think, Believe, Feel, and Want.* "Asking them why they are thinking or feeling something is more complicated because people might not know. But there's no better judge of what's going on in your mind than you."

But for those Enneagram devotees out there, a question remains: Why does it still sometimes just *feel right* to those who take it? The history of personality testing contains a clue. The first personality test was developed during World War I to assess which soldiers could be vulnerable to "shell shock," or nervous breakdown during battle.[55] But tests proliferated and boomed commercially during and after World War II, sold to organizations that wanted to better match people with jobs and increase productivity.[56] When psychologist Bertram Forer began scrutinizing the tests, he noticed they contained statements so broad that they could apply to anyone.[57] But would anyone really believe them? He tested his assumption by giving his students a personality test and asking them to rate the accuracy of what he said would be personalized results.[58] What they didn't know was he'd given them all the same results write-up, drawn from an astrology book. It included statements such as "security is one of your major goals in life" and "disciplined and self-controlled outside, you tend to be worrisome and insecure on the inside." The unknowing students rated their identical results with an average of 4.26 accuracy on a scale of 0 (poor) to 5 (perfect). Over the next few decades, Forer's results were replicated by other psychologists. In 1956, clinical psychologist Paul Meehl coined a term to describe our tendency to believe that general, universally applicable statements are specific to us: the Barnum effect.[59] Named after the circus master P. T. Barnum, the term references the idea of human gullibility, often linked to the phrase: "A sucker is born every minute."[60] I would add that you don't necessarily have to be a sucker to prefer to avoid living with uncertainty about yourself and those around you—you're just being human.

Resisting Seduction

So far, we've been operating with the assumption that uncertainty is somehow inherently and universally aversive. But for many people, this isn't true—or at least it isn't true all the time.

Even Arie Kruglanski, the psychologist who developed the need for cognitive closure scale, has revised his perspective on uncertainty in recent years. In the psychological literature, people talk about uncertainty "as if it was something that is intrinsically fear-arousing . . . and as if, intrinsically, we were drawn to certainty." *But what if neither idea is true?* he wondered. Today, he views uncertainty as "neither fear-arousing nor euphoria-activating." As with a Rorschach ink blot, it's what you project onto it—"the demons of your soul, or your inherent optimism"—that generates the meaning of uncertainty for you.[61]

What if you're one of those people, like Kruglanski and myself, for whom optimism isn't inherent? In the 1990s and early 2000s, Western psychology responded to this problem with the positive psychology movement, a trove of research aimed to help people cultivate more positivity through, for example, gratitude, forgiveness, and a growth mindset.[62] The idea was to help people learn how to approach uncertainty without fear, to "expect good things," Kruglanski says. Eastern philosophy and religion—Buddhist teachings, for instance—have for centuries suggested a different approach: We must also reduce our desire for certain things to happen, to liberate ourselves from the tyranny of outcomes. When we take away our wish for a particular outcome or answer, we can look at uncertainty with equanimity, taming it by caring less.

For his part, Kruglanski dabbles in both approaches. But most of all, perhaps, our capacity to change how we approach our big, heartwood-style questions is rooted in deeper self-awareness. To recall our question framework, these are the questions that, like the heartwood of a tree, stay with us across our lives, giving us strength and stability as we grow. "Understanding ourselves is the first step to being able to regulate ourselves, to being able to manage ourselves," he tells me.[63]

Kruglanski knows that he's "a high need-for-closure person"; he doesn't enjoy existing in a state of uncertainty and is eager to eliminate it. He's not sure whether this tendency is genetic or rooted in his own childhood trauma as a Holocaust survivor. Regardless, his knowledge of this tendency is what helps him to, on occasion, transcend it.

Even those of us with Kruglanski's tendencies are willing to explore new ideas and answers in certain contexts: when we feel secure, when we feel we have time, when we feel positively connected with and seen by others. It's when we feel threatened, isolated, and insecure that we double down on what's known.

———

Answers can give us the illusion of security and control: a sense that we are safe, that the world is known to us, that we have found meaning and coherence. But answers also provide the illusion of something else—that we have the freedom to move forward. While we can often feel stuck as we face innumerable future possibilities, eliminating all but one can, paradoxically, make us feel less inhibited.

An answer can feel like a ticket to proceed to the *next* question, the *next* answer. Getting an answer can give us a sense of momentum, a feeling that we are progressing and growing.

And, perhaps, it gives us the feeling that we are keeping up with the pace of modern life. As one theoretical physicist told me, part of our craving for answers is born from a desire to move faster. It's uncomfortable for us to sit still; it feels like we aren't advancing.

Sitting still is not a strength of mine; movement is how I trick myself into thinking I'm in control. The faster I speed through life and the more I accomplish through that speed, the more I feel as if I'm staving off death. Forcing myself into a state of constant doing, whether I'm exercising or working, allows me to avoid the uncomfortable questions that surface when I'm at rest: Who am I without the things I do? How can I appreciate my value outside the daily flurry of productive activity? How can I accept my mortality? Exercising, in particular, is connected to this last question. It's not that this connection is totally irrational; research does link exercise to increased longevity.[64] But exercising a certain amount does not mean *for certain* that I will become an illness-free centenarian.

In everyday life, this compulsion to exercise and "be productive" leads me to try to do too many things at once, prizing efficiency and speed over thoughtfulness. One morning, for instance, I was trying to bake bread for some friends and fit in a workout before an early morning Zoom meeting. In my rush, I put a new Dutch oven into the larger oven to preheat without realizing that there was still a small paper booklet inside it.

A few minutes later, while I was mid-sprint on the Peloton, the apartment began to fill with smoke and the fire alarm sounded. When we discovered the source, my husband was furious. How could I not have taken a minute to check and clean out the Dutch oven before using it?

Like adhering to the guidance of an astrologer or a psychic, chasing after the vague dicta of personal development gurus, or adapting oneself to fit a personality test categorization, speeding through life was a way of dealing with the discomfort of my questions. Moving quickly through my life not only made me more susceptible to fast, easy answers—after all, the more sped up *we* are, the more we expect our answers to emerge quickly—but also allowed me to avoid these questions about mortality and self-worth altogether, by instead immersing myself in activity.

Wrapping Up

We've learned in the previous pages about the complex reasons we can be seduced by fast, easy answers: a mix of the way our brains are structured, our personalities, our childhoods, how threatened we may feel in the moment, and who might be responding to us. In short, our response to our aversion to uncertainty may be expected, but it is not inevitable and certainly not unchangeable. But there's another layer to this discussion: It hasn't always felt this hard to deal with uncertainty. The struggle is both new and old. The way I learned this lesson almost killed me.

2

The Death (and Rebirth) of Patience

THE BIG IDEA

It's always been difficult to live with uncertainty, but we may be living in the most difficult time to do it. The little-known history and science of patience puts our current struggles into perspective and reveals an essential tool to help us break free from the paralysis of uncertainty and into new experiences, learning, and growth.

Lou and Rilke. Rilke and Lou. For a time they are inseparable, until Lou realizes she can't bear the burden of being his lover and also a kind of maternal support. Still, they remain connected, and Rilke continues to write her letters as he travels to an artist's colony in Worpswede, Germany. It's a place of "colorful, dark land under high skies constantly in motion," he writes. "Birches, tall chestnut trees, knotty fruit trees laden with red, ripe fruit." On the very first day, two "women in white" traipse up to his studio at dusk to welcome him. One is a "blond painter"—Paula Becker. The other, her friend, the "dark sculptress," is Clara Westhoff.[1]

He falls for both of them. Becker and Rilke would visit each other's studios, where they would talk and he would share his Russian books. Westhoff and Rilke walked together for hours, once chatting until 2:00 a.m. The problem is that Becker is already engaged to a man named Otto Modersohn.[2] *But this doesn't stop Becker. When Rilke leaves Worpswede for Berlin, she visits him there, attending a cooking school to help prepare her for marriage (a mandate from her mother).*

It is January 1901, and the two of them spend weeks walking together, talking, visiting galleries. Becker shares her intimate journal with him. Then Westhoff arrives suddenly, on February 3. On February 16, Rilke tells Becker that he and Westhoff are engaged. And with that, life becomes chaos. Rilke and Westhoff are heavy with doubts, worried about their uncertain future; Rilke, after all, had no money to support them. Westhoff reels from the sudden shift in their relationship. "Only two weeks ago I would have sworn it was still only a friendship," she told Becker and Modersohn.[3]

It was day 7 of a trip to paradise, and I was freaking out. The platitude "wherever you go, there you are" echoed in my brain like a toxic mantra. A. and I were in Bali, at the start of a monthlong trip through Southeast Asia with a group of friends. The trip began as advertised: The entrancing, dissonant sounds of gamelan music trailing us. Tender, spicy beef rendang—a local rice dish wrapped in a triangular banana-leaf pouch. Monkeys pilfering our sunscreen on the beach. Visits to ornate temples carved into seaside cliffs or buried in caves. But a few days in, on the island of Nusa Penida, I punctured both this illusion of perfection and several layers of skin.

One popular way to get around Nusa Penida is by motor scooter. I had never ridden a motor scooter before and begged off piloting one myself, instead hopping on the back seat of a more practiced friend. As a passenger, I gripped the back of it so tightly that my hands turned white. Meanwhile, I attempted to enjoy the azure water and jungle terraces speeding past me.

Why would I turn down the opportunity to pilot my own scooter, you ask? Here, I need to disclose something embarrassing. I have a driving phobia. The technical term for this is amaxophobia.[4]

Let's travel back in time a couple of decades. Like many teenagers, I couldn't *wait* to get behind the wheel of our family's red Volvo, anticipating the freedom of being able to drive to the mall *by myself* and to friends' houses *by myself* and to Shake Shack *by myself*... you get it.

But also, like many teenagers, I was not the most cautious of drivers. A couple of years after I got my license, I got into my first small accident on the highway driving home from my summer job. I was trying to change lanes in the middle of traffic and didn't see when the other car in front of me came to an abrupt stop. I rear-ended it, bashing our Volvo's bumper and front lights. Though no one was hurt and the damage to the other person's car was minimal, I used a summer's worth of earnings to pay for the repairs.

More important, I paid for the accident psychologically. I started to feel nervous when I drove, especially on the highway. More and more, I avoided driving, choosing public transportation when I could.

Then, during a snowstorm that following winter, my car skidded on an ice patch and I rear-ended another vehicle. Again, we were both okay, but this was strike two.

Strikes three and then four and five came next; for the curious, I reversed into a passing vehicle inside a parking lot, then rear-ended someone on the highway, and finally scraped a large cement pillar in a parking garage while backing out of a spot. Suffice it to say that by the time I was in my mid-twenties, I was telling myself a powerful story: that I was a bad driver and should avoid driving at all costs. And I did, choosing to live in places where I didn't need to drive or could rely on others for rides.

So, when I arrived in Nusa Penida, there was no question that I would opt out of scooter driving, too. But on the way back from one beach visit, as the setting sun bathed the island in gold and pink, one of my friends encouraged me to try driving myself. *Just for a minute!* she urged me. It was a rural road with no one else in sight—a great place to practice. Getting out of your comfort zone was what traveling was about, I thought to myself. It couldn't hurt to try. This wasn't an icy Chicago street or a busy city highway, after all. I'd just go slow.

I mounted the scooter, started it up, and began my trip down the gravel road, slowly at first. This wasn't so bad! It was kind of fun! I accelerated a little more. My friends cheered and waved me on. "You're doing great!" they yelled from the next scooter over.

Then, I saw a patch of gravel fast approaching at the next turn and panicked. *Wasn't I supposed to do something differently with the brake when it came to gravel?* My friend had told me, but now I couldn't remember. The adrenaline rush seemed to erase the lesson.

Whatever I did, it was wrong. I flew off the scooter and tumbled onto the gravel. Standing up shakily and brushing stones from the knee and elbow that had broken my fall, I saw the grimaces on my friends' faces.

"That does not look good," one said, pointing to the blood starting to ooze from my arm. With a makeshift bandage made of napkins from a nearby *warung*,* I returned to our beachside villa with deep wounds on my elbow and scrapes on my legs and knees. As A. helped me clean and dress them, I realized this meant I could no longer take a planned scuba dive the next day or swim in the pool. It also meant I was scared to get on the back of a scooter again and angry at myself for panicking at the sight of gravel. For the next few days, I was forced to *take it slow*, which was about as easy for me as it is for a puppy. It felt as if I was being restrained by an invisible leash.

After two days of *taking it slow*, we traveled—by boat and car—to our next Balinese location: the rice fields of Ubud. We were staying in teakwood Joglos—traditional Javanese homes with roofs that look like mountains, nestled in the rice paddies.[5] In our open-air house, we fell asleep to the orchestral arrangements of crickets and geckos. In the morning, we ate banana pancakes, pineapple, and papaya under the overhang of the outdoor kitchen pavilion, watching rain drench the verdant rice terraces. Despite the transcendently beautiful landscape, I was starting to feel a familiar anxiety. It had been days since I'd exercised or moved much at all.

"I think I'm going to go to a yoga class," I told my friends one morning at breakfast, scrolling through my phone. I'd found one about two miles away that started in a little less than an hour but reasoned that if I ran part of the way, I would just make it.

---

* The term for a small, often family-run shop in Bali.

"Are you sure it's a good idea to go in the rain?" A. asked me, gesturing at the monsoon-like storm. "Why don't you just read and relax instead?" He stared at me knowingly, an edge to his voice. He, out of all of our friends, was most familiar with my exercise and movement compulsion—how I turned to activity to avoid questions about mortality and self-worth that made me anxious and uncomfortable, and how this tendency, combined with my need for speed, could lead to dangerous situations (such as the Peloton and the pot-in-the-oven fiasco).

"It'll be fine," I said breezily, securing my bandaged elbow and knee, donning my plastic rain parka, and waving goodbye to our friends. A quiet voice in the back of my head agreed that he was probably right, but a much stronger, more anxious one reminded me that staying still was unhealthy—that I *needed* to move, that A. was simply being overly cautious. The next few days were packed with tourism activities, I reasoned. I might not get another chance to exercise.

"I'll be back in a couple of hours," I assured him.

I jogged carefully over the slippery stone path that laced through the paddies, leading me to the main street of town. This part of Ubud was a tourist mecca, the holy place for all aspiring to eat, pray, love, and otherwise find "spiritual enlightenment." Despite the rain, the narrow sidewalks teemed with soaked tourists. I was following the GPS on my phone, continually checking it to make sure I was heading the right way.

I peered at the time. This was taking longer than I'd expected. At this rate, I might be a few minutes late. *What if they won't let me into the class?* Inside my frantic mind, hijacked by the exercise compulsion, this was a worst-case scenario. I had to move faster!

I began crossing a busy street, half looking at my phone and half at the crossing.

And then, suddenly, an oncoming scooter hit me at thirty miles per hour. It launched me into the air and I hit the ground, splayed in the middle of the road, first numb with shock and then feeling an intense throbbing in my left side. The impact had knocked my phone from my hands, and it lay shattered nearby. Locals gathered to help us up. It looked like the person with whom I'd collided had only a few scratches, and her scooter was okay. My leg was in pain but I could still walk; my knee was cut up, though this time I'd injured the other one.

The person that I least wanted to tell about this incident was A. I was ashamed and shaken up. I feared his reaction. I was angry at myself. As I limped back to see him and told him what had happened, he did something I didn't expect. He hugged me, tightly. And said, "I'm sorry that this happened to you. But this scares me. And I don't know what it's going to take to make you see how serious this exercise compulsion is."

But I did see. The lesson was clear, and the stakes were high: If I didn't learn to slow down, to be patient, to take in the uncertainty of the present, I could seriously harm myself or someone else.

Who Needs Patience When You Have Smartphones?

Rilke didn't just tell Kappus to love and live the questions. He told him to *be patient* with them. This is a theme that emerges across his letters. In the third letter, he tells Kappus, "I learn this every day, learn it through pain I am grateful for: *Patience is everything!*"[6] Patience comes up again in the sixth letter, and the eighth: "Dear Mr. Kappus, there is so much happening inside you right now; you must have patience, like someone sick, and have faith, like someone recovering, for perhaps you are both. And more than that: You are also the doctor watching over you."[7] In letter nine, he writes that he hopes Kappus "can find in [him]self enough patience to endure" the many doubts and conflicts of his life, encouraging him to have "more trust in what's difficult . . ."[8]

Questions that take time for us to answer, or that are unanswerable, can be some of the most difficult and rewarding parts of our lives. If you think back to the parts of a tree that symbolize different kinds of questions, we need patience to wait for certain fruits, like the pawpaw, to ripen—or to sit with questions that may never be permanently answerable, staying with us throughout our lives like heartwood. Patience is inextricably linked with our ability not just to live *with* uncertainty but to live *better* in it. Both the word for the virtue of patience and the label of someone as a "patient" originate in the Latin root *pati*, which means "to suffer," and *ferre*, which means "to bear."[9] Patience is about waiting well, waiting calmly, while we suffer. It's about bearing reality without giving up on it. On the journey to love the questions, becoming more patient is one of the first and most essential steps.

Not many of us would try to refute the idea that patience is generally a good thing. But if you're like me, the idea of intentionally learning to be more patient sounds about as fun as inching along in rush-hour traffic or enduring a sixteen-hour flight with a crying baby to soothe. Though patience is a struggle for so many of us, it's also not high on the list of personal growth priorities. Who wants to practice suffering well?

Uh, no thanks.

It turns out that patience has been out of style for decades. And learning about this history helps us understand why the quest for patience is a particularly challenging one today. In his book *Patience*, scholar and former president of Allegheny College David Baily Harned argues that patience fell out of fashion, both in Christian texts and the secular world, with the technological and social advances of the Industrial Revolution.[10]

By 1848, signals of change such as the publication of Karl Marx and Friedrich Engels's *The Communist Manifesto*, uprisings throughout Europe, and a quickening speed of innovation all "conspired in the discrediting of patience," Harned writes. "The virtue came to be seen either as an anachronism or as a notion devised by oppressors to contain the restlessness and discontent of the oppressed."[11] The act of patiently enduring adversity was perceived, not as an opportunity to become spiritually stronger, but as "a diminishing of the quality of our lives, a deprivation enforced upon us by an unfriendly environment . . . Far from being one of the great human acts, waiting simply testifies that we have not won our struggle with the world."[12]

Fast-forward to the development of the internet and then the advent of the smartphone, a tool that contributed to the belief that avoiding long periods of waiting or uncertainty is progress. More than a decade ago, this was the conclusion of a group of academic scholars out at a pub in Saskatchewan, Canada, near the University of Regina. They'd been debating the answer to some question that University of Regina psychology professor Nicholas Carleton can no longer recall.[13] Earlier in the night, he said, it might have been about the effect size in an experiment. A little later, with empty pints crowding the table, it likely would have been about which actor played a character in a movie. Whatever the question was, it prompted one member of the group to remark how nice it was that they could avoid debating about the right answer and could simply pull it up instantaneously on their devices.

"That *was* terrific, but it was also potentially a problem," Carleton recalled thinking at the time. He's spent his career seeking to understand the factors that increase risk for mental health issues and how those factors change over time.[14] The problem, he suspected, lay dormant in many people's new relationship with questions and answers. This new relationship had unintended consequences. With each answer smartphone users looked up, they missed out on an opportunity to "sit with not knowing in relatively safe environments," he told me. "We know from mental health disorder treatments that exposure-based therapies can be very powerful to help people overcome anxiety. And a big part of anxiety management can be learning to tolerate uncertainty, to become comfortable with not knowing. There I am with a cell phone, and all of a sudden, I don't need to tolerate that. I now have a device that can build in certainty, or perceptions of it, into every part of my life, taking away opportunities to practice tolerating uncertainty."

Carleton became curious about how widespread smartphone use might be influencing our collective capacity to tolerate uncertainty. He analyzed self-reported levels of intolerance of uncertainty, or IU, between 1999 and 2014, alongside the penetration of mobile phones and the internet. In 2018, he and his colleagues published research that suggests we are becoming *less* tolerant of uncertainty over time; this reduced tolerance correlates positively with the spread of devices.[15] Glittering with answers, our smartphones can increase our tendency to seek reassurance, acting as "safety cues" and reducing the spontaneous, everyday exposures to uncertainty that might help us develop a stronger tolerance for it. So, the proverbial muscle starts to atrophy, increasing our *intolerance* of uncertainty and putting us at greater risk of anxiety and related disorders.

Of course, as any student of statistics knows, correlation does not equal causation. In other words, just because these variables seem to be related does not mean one is causing the other. Without running a controlled experiment, we can't say that my increased smartphone use is *causing* me to be less tolerant of uncertainty. Indeed, Carleton readily acknowledged that smartphone use is just one piece of the puzzle of explaining the overall trend.

"Part of it is instant access to information, which has changed dramatically in the past twenty, thirty, forty years," he explained, including the internet and the spread of personal computers as part of this trend. "This has provided more opportunities for us to be scared by the incredible uncertainty that exists out there."

The climate crisis is just one of many ambiguous, uncertain threats we now face.[16] We know it could vastly alter or possibly end civilization, but we also can't fully grasp how imminent the threat is or how much control any one of us has over it. If we try to find guidance online, we can quickly experience information overload.

Carleton cited an overall increase in mental health disorders, which he attributes to superior tracking systems and our uncertain times.[17] "New disorders are coming out," he said. "More people are saying they don't want to have children because they are terrified of the future," citing climate change and economic uncertainty.

On the one hand, we have more answers at our fingertips than ever before. On the other, we have access to more information than ever about how uncertain the world is. Our smartphones are not designed to help us deal with this dilemma. But the art and science of patience just might be the key to unlocking a solution.

Uncovering the Science of Patience

There's a ubiquitous assumption of our "impatient age," Harned writes.[18] That's the assumption that *real life* is characterized by action, agency, and productivity—that we are most fully human when we are in control and *doing things*.

Sarah Schnitker was a first-semester PhD student studying personality and social psychology at the University of California, Davis, when she picked up a copy of David Baily Harned's book to read over winter break and began reflecting on her own relationship to patience and productivity. Ever the model grad student, she was "trying to read all of the literature" as she considered her career focus.[19] This was 2004, and the science of character strength development was beginning to gain traction.[20] But Schnitker "was struck that no one was talking about patience—at all." Indeed, throughout the twentieth century, only one researcher—the University of California psychology professor Albert Mehrabian—attempted to measure patience empirically.[21]

As she continued to read the handful of patience manifestos, written largely by philosophers and theologians, she became more convinced of the role patience could play in human development, as something that could help us lead fuller and more meaningful lives.

"I wondered, why aren't we studying this in psychology?" she recalled.

Patience, she learned, was a misunderstood virtue—both in academia and in the larger world. When people think about patience, "they think you've given up or you have become a doormat. You've become passive, and that's so against our culture."[22]

Aristotle describes patience as the golden mean[23]—or the ideal way to behave morally—between two vices: recklessness (a.k.a. impatience) and sloth.[24] Schnitker suspected that patience could have more to offer. The place to start was to figure out whether it was a distinct trait and whether it could be measured. With Robert Emmons, a UC Davis psychology professor, Schnitker developed a scale to measure individual differences in the trait of patience and tested its validity with UC Davis undergraduates.[25] She and Emmons found that the scale was valid and that patience was both distinct from other traits and significantly related to experiences of spiritual transcendence, as well as other kinds of religious behaviors.

In 2007, Schnitker and Emmons published their paper introducing their new way to measure patience and their results. This was a huge step in Schnitker's career. As she received encouraging feedback on the paper, she knew she was onto something.

And then, suddenly, the promise of continuing her career ascent slipped away.

Shortly after publishing her paper, in the early summer, Schnitker woke up one night throwing up. At first, this didn't seem too unusual. Though she wasn't running a fever, she figured she had a stomach bug or some virus that would run its course. But for nearly two months, every few days, she would vomit unpredictably, not knowing what had triggered the vomiting. She became a regular visitor to urgent care to get intravenous fluids because she could barely hold anything down.

Doctors were baffled. First, they suggested an endoscopy, which showed nothing unusual. Next, she got a brain scan that, again, turned up nothing. Many suggested the cause was psychological. One doctor prescribed a drug that made her suicidal.

In six weeks, she shed thirty pounds from her already trim five-ten frame. Her condition converted every activity into a potential risk, tinged with dangerous uncertainty. Could she manage to go to the grocery store without throwing up? What about driving? She had two fender benders in a week because she was vomiting uncontrollably in the car.

For months, she wondered whether she would ever overcome this mysterious affliction. She began to get fevers. "It was as horrible as it sounds," she told me. "That was the most difficult year of my life."

Though she had been living alone, she no longer felt safe on her own. She moved in with a family from her church.

In the meantime, Schnitker was still a graduate student in the middle of her program. She was "supposed to be jumping into my research with both feet," but she never knew when she would feel sick. Days and then weeks disappeared from her timeline. "I had to learn how to continue, how to not fall apart and have all of my life goals disappear," she recalled. "I had to be patient and self-regulate, to recognize 'my body won't do this today, I can't push it or I'll make it worse.'"[26]

Ever the scientist, Schnitker began to get curious about her own experience. Far from inducing more passivity, she found drawing on patience was helping her maintain course toward her PhD objectives. She began to wonder: Could patience influence someone's pursuit of goals—allowing them to lead more rich, vibrant, and purposeful lives? If patience was a superior approach during periods of uncertainty or suffering, how did it impact someone's sense of well-being?

After six months, doctors finally delivered a diagnosis: cyclic vomiting syndrome.[27] But, as with many chronic or autoimmune illnesses, the origins and treatments remain mysterious. "No one knows why it happens," Schnitker said. After she was finally diagnosed, it took another year and a half before Schnitker felt that the condition was at least somewhat under control. But that sense of control is an illusion. Even today, she said, under conditions of stress or with weather shifts, the vomiting can return. "It's still something I struggle with," she said. "But learning in my twenties that I don't have full control over everything has been a gift. I hate that I have this, but it also allows me to release control of other things." She couldn't control her illness, but she could control how she moved with and through it—not allowing the uncertainty it created in her life to keep her stuck.

As Schnitker learned to live with her illness, she continued to study patience, and she designed her next set of studies with UC Davis undergraduates to answer some of the questions prompted by her own experience of patient goal pursuit.[28] In one study, participants reported ten goals, and researchers tracked their goal achievement and patience regarding their goals across time (at five periods across a ten-week academic quarter). Students rated how much patience they thought each goal required and how much patience they personally had about achieving it. Researchers also surveyed "informants" who knew study participants personally. These surveys asked about the informants' perceptions of the study participant's patience. Self-reported data is notorious for its tendency to be biased; adding reports from knowledgeable others can be a method to introduce more objectivity to an analysis.

Based in part on her own experience during her illness, Schnitker and other researchers hypothesized that the more patient someone was, the more effort they would exert in trying to achieve their goals. She and others also believed that exerting more effort toward achieving goals would yield greater progress, which would make people feel better overall. Patience is a tool that "helps [people] stay emotionally regulated as they navigate uncertainty, waiting, and suffering," she said.[29]

When someone is patient in pursuit of a goal—such as trying to find the answer to a question—they continue to be aware of and attentive to that goal even when the pursuit grows more challenging and they may crave distraction. She discovered that, far from being passive or weak, patience is an essential ingredient not only to achieving goals but to living well and actively during periods of uncertainty.[30] Her work has shown that higher levels of patience are correlated with decreased depression, decreased negative feelings, and lower incidences of health problems.[31]

Even as evidence piles up, Schnitker recognizes that patience still has a PR problem. Practicing patience doesn't sound fun or important, particularly in a society that devalues it. "The way I sell practicing patience is [to tell people]—you know you'll have a time in the future when you will experience suffering," says Schnitker. "Do you want it to be a soul-crushing thing that destroys you and what you care about, or do you want to be prepared so that you can still pursue the things you value, and move through it well?"

In other words, do you want your relationship to uncertainty to paralyze you or propel you toward a stronger, more dynamic life?

"Patience is not resignation, passivity, or inaction; rather it is the emergence of freedom within the domain where necessity rules," Harned writes.[32] One way of interpreting his words is that the anxiety we feel about our biggest questions has the power to rule our lives—to hijack our energy and attention.

In contrast, patience "is how you say, *I'm choosing to engage with this uncertainty in a certain way*," Schnitker tells me. "The paradox of patience is that it's about being receptive, but also choosing to receive [the experiences the world is giving you]. By acknowledging the reality that you aren't in control, by embracing the questions of your life, you regain your freedom, agency, and autonomy."[33] In other words, the questions of your life no longer control you when you *choose* to patiently explore them, rather than feeling like a victim of the uncertainty they contain.

For Schnitker, who today leads the Science of Virtues Lab at Baylor University in Waco, Texas, it has also meant asking herself big questions: *"How do I navigate making a meaningful life even with this disability? And how can I define myself, and know my worth, outside of my actions and accomplishments?"*

For my entire life, I'd attached my own sense of worth to my accomplishments, allowing external appraisals of me to determine who I should be. The love I felt for myself was conditional. If I didn't meet a set of exacting standards, my inner critic lashed out ferociously, telling me to do more and try harder. As long as I continued striving, I knew who I was and where I was going in my life. Stopping would mean sitting with those uncomfortable questions. I wondered what it would take for me to see myself as valuable without continuous accomplishment and striving. And I wondered how Schnitker had done it.

As I talked to her, I became curious about how Schnitker's approach to patience showed up not just in her professional life but also in her personal life. It's one thing to be patient in pursuit of a PhD or another concrete, linear professional goal. What about the pursuit of richer relationships? What about the pursuit of appreciating what life already holds? It's easy to talk about these ideas, but what do they look and feel like? And could Schnitker use her expertise to help me?

To find out, I knew I had to visit her in Texas.

Is Patience Bigger in Texas?

It was not easy to get to Waco from California in early January. At several points, it felt as though the trip was specifically designed to test my patience and capacity to deal with uncertainty. First, I was called for jury duty the week before my scheduled visit. Then, Schnitker also received a jury duty notice. A few days before the trip, we were both released (*phew*).

Then, on the day of departure, the Federal Aviation Administration grounded all flights in the morning because of a technical issue. In the Bay Area, stormy weather caused further delays. Miraculously, my plane eventually took off, though I had to sprint through the next airport to catch my connecting flight. *I'll make it after all!* I thought happily as I munched my free pretzels.

But there was one final hurdle to clear. When I reached the Dallas Fort Worth International Airport later that evening, I realized after a few minutes that I'd made a naive and incorrect assumption: that it would be easy to get a ride share from there to Waco. Driver after driver pulled up and canceled the ride as soon as they heard where I was trying to go. (For reference, Waco is about 110 miles from the Dallas airport, or about an hour and forty minutes without any traffic).

"Need a ride?" a bearded middle-aged man asked, about thirty minutes into my cancellation journey. "My cousin is coming to get me, and we could take you somewhere."

"No, thanks," I said, walking away quickly and realizing that I needed a plan B. The only viable option I could think of, other than waiting around for a ride I might not get, was to rent a car and drive myself to Waco. As I hopped on the bus to the rental car lot, I anxiously considered the variables that would make this trip especially daunting: I'd be in a different car, in a new place, at night, on a highway. For someone with my level of phobia, this was some nightmare version of the driving Olympics. I wished I was one of those people for whom getting behind the wheel was no big deal—an easy, mundane part of life.

At the rental car counter, I requested a sedan—something safe and affordable. "Sorry," the clerk said, "we're all out of sedans. The best I can do is to get you one of our luxury vehicles." Great. I hadn't budgeted for a car, let alone a luxury car rental. Finally, around 9:00 p.m., I was gripping the steering wheel of a BMW 5 Series, singing loudly to an assortment of '90s hits and musical soundtracks, and praying that I'd make it.

When I pulled into the parking lot of my hotel, I felt an embarrassing sense of pride—as if I had just, through some feat of extraordinary cunning, narrowly avoided death. I remembered a trite quote that a friend had shared recently on social media: "Feel the fear and do it anyway!"

Despite being cliché, this phrase pointed to an important question about exercising patience: How do you know when to act? How patient is *too* patient? In my case, waiting endlessly at the airport for a ride would have been foolish, even though I was nervous about driving myself. Indeed, patience is not by itself a sufficient tool for pursuing goals and breaking free from the paralysis of uncertainty. Sometimes, we need courage.

The importance of cultivating courage alongside patience is borne out by science. According to researchers, courage is the virtuous sibling of patience; they counterbalance each other, encouraging us to stay somewhere between reactive recklessness and disengaged passivity.[34] Patience and courage both help us better regulate our emotions in the uncertain pursuit of goals and in the murky space of unanswered questions.

Now, it's a probably a stretch to classify my decision to drive as demonstrating capital-*C* courage, though it does seem to fit the word's definition: the ability to face obstacles or fears and continue despite them. But as I texted A. excitedly from my room, "I MADE IT!!!!!"* I was too relieved to care.

Many Americans associate Waco with the 1993 siege, a fifty-one-day standoff between a religious cult—the Branch Davidians—and US federal agents.[35] Today, it's a city trying to build its artistic, culinary, and aesthetic character. Here, a patchwork of barren lots and low-income housing sits alongside sleek new hotels and hipster coffee shops. The downtown is speckled with bright murals depicting historical scenes and touristy slogans, including one that reads "City with a Soul."[36] On my way to Baylor's campus, I passed whimsical restaurants and bars—one featured a series of swings on a porch designed for social media photos. I saw an island of fast-food chains and road construction everywhere. Waco is not very walkable, and indeed, few people crossed my path. As I wandered down a sidewalk, a driver stopped and yelled at me, "Do you need a ride?" Even though this was a much less shady character than the guy at the airport, I declined.

The Baylor Sciences Building loomed over the Waco Creek. It opened into a bright, empty atrium dotted with plants, a shuttered Starbucks, and a Mexican take-out spot. Most of the students were still on holiday break. Schnitker's office was on the third floor, down the hall from another professor with a sign on his door that read, "A scientist is not someone who gives an answer, but who asks the right question."

"He just put that sign up recently!" Schnitker told me. "I thought of you when I saw it."

The first thing I noticed about Schnitker was that she appeared to vibrate, her exuberance rippling through the air.

"Once I got kicked out of church for being too loud," she told me, recounting the story of how, when a church member complained about

---

* Yes, I used that many exclamation points.

her too-boisterous singing voice, Schnitker was asked to either quiet down or leave. She found a new church.[37]

Her office was cozy, with a plush purple couch, a bowl of lollipops on her desk, family photos, and the requisite scholarly wall of books. On the wall was a poster from a visit to Istanbul—a favorite trip she took with her husband—and a painting that looked like, well, nothing when I first saw it. But Schnitker encouraged me to look again, and then I could see it: the shape of a woman, subtle and emergent.

Also on display were two mentoring awards she received from student nominations. Mentoring graduate students and postdocs was what she loved most about her job, she told me.

And this was something her postdocs loved most about her, too. Over the course of my visit, several told me, unprompted, about Schnitker's mentorship—her patience with them and her courage on behalf of their growth. "She fights for grad students," a couple of students shared with me. "She won't stand for mistreatment."[38] Her lab was in the top five of grant recipients in the entire Psychology and Neuroscience Department, having secured $10 million as of early 2023.

"The culture of the lab is very intentional," said Juliette Ratchford, a postdoctoral fellow with a passion for Dungeons & Dragons. "Everyone holds each other up, which is rare in academia. In other places there's a fear-based model—do things so you don't fail. In our lab, Sarah celebrates failure because it's a huge part of academia."[39] Each week, the team would meet to talk about progress they'd made on different research projects by moving sticky notes across a board divided into different phases of academic publishing—for example, "writing" and "submitted." On the board, "rejection" counted as a success, because it's part of the process and helps the researcher move on to other potential opportunities.

In Schnitker's lab, emphasizing self-worth outside of academic achievement wasn't just a cultural norm, it was also a subject of investigation for many of the people working there—like doctoral student Elizabeth Bounds.

Growing up as an athlete, Bounds became curious about the questions "Who am I, and what am I worth if I can't achieve and perform?" As she played soccer and basketball, Bounds began to realize her self-worth was too rooted in her success on the field or the court. "For so many athletes, sport is a very salient source of identity," she told me. We were sitting with Jay Medenwaldt, then a PhD candidate in social psychology (and the one who did the Enneagram analysis in chapter 1), in the plant-filled atrium of the science building.[40]

The problem with any kind of achievement or performance-related identity, Bounds told me, is that sometimes it subsumes you entirely. Being an athlete is not just part of who you are, but all of it. Also called identity foreclosure,[41] this problem can make it seem as though a failure on the playing field indicates that you are a failure. It turns any loss, injury, or mistake into a kind of referendum on your value and dignity. "What I know now is that it's more adaptive to root your sense of self-worth in more sturdy, internal domains—for instance, in your own morality," she said. "Work performance, athletic performance, other people's opinions—all of these are outside of your control. Those things end up burning you."

Today, she's interested in how cultivating both patience and courage can help people face threats to their self-worth. "I'm interested in courage because I think doing anything meaningful ends up being challenging or scary at some point," she said. "Courage helps you do hard and meaningful things."

"That reminds me of a recent meme I saw," said Medenwaldt. "It said—we don't do these things because they're easy. We do them because we *thought* they would be easy." We all laughed.

At the end of the day, I'd learned a lot from observing Schnitker and her lab. But that was only one part of the reason I'd come to see her. My compulsion to move and inability to slow down was fraying the most important relationships in my life. I wanted to see how her approach to patience showed up with her family—and to take lessons back to mine.

The Professor of Patience at Home

Later that evening, as the sun set and nearly blinded me on the highway, I drove to Schnitker's house. I approached the redbrick, colonial-style edifice, the entrance fortified with white stone pillars. As soon as I rang the doorbell, I heard a dog bounding across the floor. A little girl squealed, followed by Schnitker's muffled voice: "Do you want to open the door?"

Emory, Schnitker's six-year-old daughter, swung open the door to their home. Their one-year-old black-and-white labradoodle, Holly, jumped on me, licking my face. We made our way from the foyer to the kitchen, where Emory immediately began to show me her rock collection.

As we sat down to dinner—salmon, blistered green beans, couscous, and a spinach salad—the family said grace, and Emory, like most six-year-olds, eagerly interjected her questions and stories into moments of attempted adult conversation.

"Can I say something?" Emory asked.

"Yes, you've been waiting patiently," Schnitker responded.

Today, she told us, someone at her school had to eat a still-frozen waffle. A frozen waffle! This was a mind-blowing event, and she giggled as she recalled the absurdity. She also asked whether we'd like some of the dessert she'd made that afternoon with her grandmother. To me, it looked like puppy chow—the concoction of Chex cereal, chocolate, peanut butter, and powdered sugar that was also a feature of my youth. "No, I renamed it!" Emory informed me. "It's called 'A Snowy Day.'"

We finished dinner and headed to the living room, settling into couches that faced the mantel, where Christmas stockings were still hanging; the tree was still up, too. "Can I show her my pajamas?" Emory asked Schnitker, gesturing at me shyly. And then, after I agreed that I would like to see her pajamas, she confirmed, "Mom, I like her!" Being endorsed by a six-year-old felt validating.

In her office the next day, Schnitker told me that motherhood supplies her with the most regular tests of patience, second only to her chronic illness. "Every dinner, it's a practice of patience for the whole family," she said. "I make Emory wait to talk, and I try not to get frustrated. You have these small opportunities every day to practice so when you're dealing with something big, you can handle it."[42]

And with that practice, Schnitker made a discovery: If she does manage to stay calm amid her daughter's interruptions, or tears, or constant questions, she receives something in return.

"Especially in these young ages, when kids show serious emotional dysregulation, it's very easy as a parent to be like, 'Stop it, I don't have time,'" Schnitker said. "But there are times where I sit with it a little longer, and ask the question 'Is there a reason you're doing this?'"

In those moments, sometimes Emory can "verbalize something that changes the whole experience—suddenly, her reaction makes sense, and our relationship is better." Patience, Schnitker said, gives her the "opportunity to see my daughter develop into a different person."

That's true of the role that patience can play in any relationship—whether it's with our questions or with another person. Patience can allow us to see a question to the end of its cycle: the ripened fruit of an answer, an emerging new question, or dead leaves we'd be better off releasing.

Making Patience Personal

Something about Schnitker's research was still nagging at me: How could I apply her findings to my own life? In one of her studies, she developed an intervention to help people increase their patience. Framed as a stress reduction study, it included group discussions, guided meditation, and other individual activities—such as considering frustrating past situations.[43] Researchers asked participants how they might be able to reframe or reappraise those experiences. The intervention increased patience in participants.

Toward the end of my visit to Waco, I asked Schnitker, "How, based on your research, would you advise *me* to strengthen my own capacity for patience?"

Schnitker thought for a moment. "First, I'd want to know, What does patience mean to you? What's driving this desire to become more patient right now?"[44]

I started by explaining how my lack of patience shows up in my relationship. "I feel very uncomfortable moving slowly," I told her. "My husband is constantly saying, 'Slow down!'"

"I get that!" Schnitker said, nodding with recognition.

"But what I've come to realize is that this tendency is something that could harm me," I said, relaying the story of the Bali motor scooter accident and how it happened in part because I had been rushing to compulsively attend a yoga class.

"It sounds like you got caught up in something you thought would regulate your emotions that day—but it probably wasn't really about the exercise," she said, as I nodded. More likely, she said, it was about avoiding what bubbled up when I slowed down—the emergence of "big questions that feel more threatening."

"Like, about death," I said, telling her about how exercise made me feel as if I were in control of my mortality.

She nodded vigorously. "Yes, death!"

I made a mental note of the fact that, in trying to avoid thinking about death, I might have almost caused my own. And that brought me to a quandary: "Even though the stakes have become high, I still notice myself doing things every day that aren't conducive to becoming more patient," I told Schnitker. In other words, I knew I needed to change, but I still wasn't changing. Why not?

"What would be the positive state if you changed?" Schnitker asked. "We know what you're trying to avoid. But who are you aiming to be, and how does that relate to your ultimate goal in life? What's important for you that we could connect the positive vision to? Identifying that is important. And you would do that by asking big questions—like 'What is my purpose here?'"

Lest you think she was simply improvising, Schnitker was following an evidence-based protocol to work with me. She was using something called motivational interviewing to help me connect the behavior I wanted to change with the person I wanted to become and the goals I wanted to achieve.[45]

"A lot of times, our goals start at a basic level—'I want to start eating healthily,' 'I want to be less compulsive,'" Schnitker explained. But it's difficult to achieve those goals unless "we have the big why. If you just have the how, you'll give up."

My "why," I told her, was twofold. A big part of my identity is being a writer and a journalist—someone who will contribute to society by sharing narratives that help people better understand themselves and the world around them. The greatest writers are also the keenest observers. They perceive things others don't, in part by moving more slowly and looking more closely—not via compulsive speed walking.

"So this really matters to you!" Schnitker exclaimed.

"It matters a lot," I said. "I'm so caught up in my anxieties that I'm skimming over my experiences. Sometimes it feels like I'm failing as a writer because I'm not being as observant as I could."

My other important goal in life, I told her, was to start a family and, even more than that, to sustain strong relationships with the people I loved. To me, this meant I needed to do more than just show up; I needed to be present for those people. I didn't want to sit in moments with my friends and family focused on the next thing, wanting to hit a fast-forward button. I didn't want my marriage to be characterized by impatience—an eagerness for my husband to change, for us to change, for questions about our relationship to be answered. Like Schnitker with her daughter, I wanted to give those around me the opportunity to grow into different people and for our relationships to evolve in unpredictable ways.

"For someone to love you, for it to truly be a loving relationship, you have to be able to receive their love," she said. "Patience is about being receptive rather than agentic."

Schnitker advised me to think about a daily practice that could help deepen my awareness of my impatient tendencies and begin to exercise my patience muscle. It would be most helpful, she said, if I could connect the practice to something I already did. For instance, before I take my first bite of breakfast, I could notice three things around me (instead of scrolling on my phone). Or, on my daily walk, I could take the first five minutes to reflect on one thing I'm going to do more slowly that day.

When she was trying to reduce her workload, she told me about one ritual she developed. "As soon as I would get into the car [at the end of the day] I would reflect: 'What's one thing I did today that I could have asked someone else to do?' It didn't have to lead to immediate action, but it was just noticing."

Cultivating patience isn't only about creating new rituals. It's about considering which obstacles might obstruct my path—also known as *mental contrasting*[46]—and coming up with a plan if those obstacles arise. What do I do if I feel tempted to fall back on my compulsive and impatient behaviors? Implementation intentions are a research-backed intervention that offer a way to address this question.[47] The idea is to create an if-then statement that helps provide you with rules and structure so you know what to do when you're in a situation that normally triggers undesirable behavior. For me, such a statement might be "If I'm feeling compelled to exercise but I don't have time, then I will do a ten-minute meditation." Your implementation intention makes clear exactly *when* an obstacle to your goals might come up, *what* you'll do, and *how* you'll do it.

As Schnitker and I leaped from the peach and pawpaw questions (the answerable ones from our tree framework) to the deeper heartwood variety (the ones that stay with you throughout your life), she encouraged me to consider these questions: What would it look like a year from now for me to be more patient? How would my writing be different? How would my marriage be different?

"You've got to be connected to that," she said. "And then the trick is to connect your practice to a community that's doing this. That's where it gets a little more challenging."

Here, we face a paradox: We're able to sustain our ability to love the questions of our lives only when we embed ourselves in the right relationships and community. But the act of entering relationships can often mean facing even more uncertainty.

"This fear of the inexplicable has done more than impoverish the life of the individual," Rilke writes. "It has also constricted relations between people, lifting that relationship out of the riverbed of infinite possibilities, as it were, onto an unseeded plot of land on the shore, where nothing is accomplished. For it is not laziness or inertia that makes human relationships repeat themselves from one case to the next, with such unspeakable monotony and stagnation—it is also our shrinking away from any new, unpredictable experience we don't feel able to handle."[48]

In some ways, the world we live in has made it easier than ever to avoid the kinds of relationships and community that are essential to our growth. Many of us live lives that are not designed around community membership—to our great detriment. Many of us now suffer from a well-documented loneliness epidemic.[49] We move frequently.[50] We spend more time alone and less time with friends.[51] Those of us who work remote jobs can find ourselves huddled over laptops for hours, watching calendars fill up with virtual meetings that may leave us at the end of the day feeling even emptier and lonelier.

And in these moments of disconnection and uncertainty, we too often look impatiently for answers in the wrong places—and rarely find what we're really looking for.

3

Seeking the Right Sources to Ask

THE BIG IDEA

Searching for answers in the wrong places, at the wrong time, and from the wrong people, can lead us into relationships with technology and other tools that don't fulfill our real needs. Often, what we're looking for is not just an answer, but genuine human connection.

Rilke's daughter, Ruth, is born on December 12, 1901, just eight days after his twenty-sixth birthday. His family promptly cuts him off from his regular stipend—meant only for Rilke the Student. He searches for opportunities to make money for months, all of them fruitless. And he grows more and more impatient, oppressed by his crying baby, the novelty of being a father now tarnished. "The walls closed around him like a prison and Rilke was restless," writes his biographer Ralph Freedman.[1]

And so was Clara, an artist who also missed her work. "I am . . . so very housebound," Clara writes to her and Rilke's old friend Paula Becker. "It is impossible for me simply to get on a bicycle and pedal away as I used to do. . . . I . . . have a house that has to be built—and built and built—and the whole world stands there around me. And it will not let me go."[2]

The marriage is disintegrating. Rilke sees no other path but to leave. And then, he receives an opportunity: to write a study of the famous sculptor Auguste Rodin. He will need to live in Paris.

He leaves Clara and Ruth behind. And that is where they stay—the recipients of his letters, but rarely his loving presence.[3]

This was the story that I had been telling myself: Before A. and I were married, during a particularly rough patch, I went to the aforementioned astrologer for entertainment. I didn't really believe in astrology; how could I when the scientific evidence shows its predictions don't hold up?[4] The astrologer told me something about my relationship that upset me, though—that we weren't a good match for each other. I balked at her certainty, subconsciously blaming her, and astrology in general, for being exploitative, one of the Charlatans of Certainty. Her voice haunted me for years, casting doubt on my choices and whether we would make it as a couple.

But . . . something nagged at me. This story was too simple.

Several years after that reading, I found Jessica Lanyadoo. A self-described astrologer, psychic medium, and animal communicator, she has been practicing astrology since 1995. I can imagine some of you raising your eyebrows as I write this. Or perhaps you're a fan of astrology who thinks that, so far, I haven't been fair to the practice. The journalist in me knew that to tell a reputable story, I had to at least talk to someone on the other side. Doing so, I reasoned, wouldn't force me to decide whether I buy the claims of astrology as valid. I would just be recognizing it as a significant part of many people's lives,[5] and trying to better understand how it fits into the modern relationship we have to uncertainty and questioning. Dismissing astrology out of hand would be taking the easy way out. Asking about what it might have to teach me, given its longevity and popularity, felt harder but important—especially because, given my reaction, astrology clearly *did* mean something to me, even if part of me wanted to pretend it didn't.

Lanyadoo and I met on Zoom; she was wearing round, neon glasses, and her curly brown hair was pulled back. She was a little late because she had accidentally entered the wrong Zoom room link.

"Mercury in retrograde, what are you going to do?" she joked—but with her, it was not really a joke.[6]

I asked her about the role she thought astrology played in our addiction to fast, easy answers. Her answer: Don't lump all astrology practitioners and practices together. Astrology is a lot of things, she said. Predictions and horoscopes, for instance, are only a small sliver. And astrology is a lot of people: Some people practice the craft responsibly, and some do not (i.e., the Charlatans of Certainty). Part of the problem, she believes, is structural and societal: There's no governing body for astrology and astrologers, the way that there is for therapy.[7] And social media has increased the amount of garbage astrology in circulation, along with the numbers of irresponsible astrologers. We also live in a society that encourages us to "power through" traumas and difficult situations, telling us that we need to find an answer or a solution as soon as possible. "But the reality of being human requires introspection, reflection, fear, loss, and sadness," Lanyadoo said. We can't just speed toward resolution, as much as we'd like to try.

We also can't simply blame external forces for our desire for answers, she said. We must take responsibility for the information sources we choose to trust and the questions we ask them. Indeed, she told me, a big part of her job when she works with clients is "helping people find the right questions. Most of the time people are asking the wrong questions. In my field of work, people come in wanting quick and easy answers; they want absolutes. It's almost always the wrong approach."

For Lanyadoo, astrology is fundamentally *not* about doling out fast, easy answers. She sees it like coaching or therapy, as another source of data and another analytic tool (among many) that people can use to help them make decisions. Astrology to her is a resource to help people develop a closer relationship to who they really are and what they really want.

"When people come to see an astrologer, they think I'm going to answer their questions," she said. "I did that in my career at first, because it was expected." But then she learned how to counsel and consult with people through the process, helping them refine their questions and motivations. This "means sometimes I don't have to answer their questions—they answer them for themselves. That's why I have a successful career, because the center of what I do isn't predictions—it's helping people to help themselves."

I told Lanyadoo the story of my experience with the astrologer several years ago. She asked me what question I asked the astrologer. In my memory, it was something like whether A. and I were a good fit or whether we would stay together. I told Lanyadoo that I'd confessed to the other astrologer that we'd been having problems and were going through a rough patch. I explained how upset and desperate for answers I'd been. And yet, I'd also told everyone, including myself, that what I was seeking was entertainment—a fun distraction.

"You made a classic mistake," she said. "You asked the wrong question of the wrong tool."

For instance, she said, let's say you have a friend you love. But she makes terrible dating decisions. Now let's also say you continually ask that friend for advice about your love life. "It's not just about asking the right questions, it's also about asking questions from the right sources," Lanyadoo said. "You went to the astrologer for entertainment, but you asked a real-life question. You went feeling desperate and hurt but thought, 'This will be fun, I'll ask all my heartfelt and vulnerable questions, it's no big deal.' Most people abandon common sense with astrology, and intellectuals do this the most because you think, 'I don't have to take this seriously because it's silly.' And then it ruins your life."

She was right. Fundamentally, I'd been lying to myself about what I really wanted from the experience and what I was looking for. I hadn't thought about it much at all. I was using the astrology consultation as if I believed I was a helpless victim of fate, rather than a responsible agent in my own life.

This problem with mismatched expectations—asking the wrong questions of the wrong tool—isn't just limited to astrology, Lanyadoo said. We also have the same problem with therapists, expecting they'll tell us what to do or give us advice. The point, she said, is not to chastise me or other people, but to acknowledge that this is part of being human. "We all ask the wrong questions and have them lead us down a painful path," she said. "And then we either learn, or we don't."

If Lanyadoo had been consulting with me, she wouldn't have peered at my partner's chart, but she would have talked to me "about what's wrong and what's not working," she said. "The key for me is in counseling people through their questions."

"What distinguishes the right question from the wrong question?" I asked her.

"The questions I get are lacking in nuance and agency," she said. "Like, 'I'm a Pisces and I'm dating a Gemini. Will it work?' That's a ridiculous question. That's not how it works. What does it even mean, will it work? Does that mean, 'Will we work forever?' 'Will we fuck forever?' That's the most common question. If someone comes to me and says, 'Will I stay with my boyfriend?' I will generally say, 'Why are you asking?'"

Counseling also means making sure the person *really* wants the answer to the question they're asking. I considered this, realizing that I *hadn't* wanted to know the astrologer's answer to the question. And the reason was that I already knew the answer I wanted to hear: that we would get past the rough patch eventually and that we could make our relationship work. The better question, I realized, was not whether or not we should be together, but *how* we could be together. What might we both have to change in order to make it work?

It was easy to blame others—the Charlatans of Certainty or even A.—for my circumstances: the relationship confusion, hunger for answers, desire to resolve it all. But blaming others was only part of my problem with loving the questions of my life. The more important part was that I lacked patience and personal responsibility. I wasn't slowing down enough to think about the question that would actually help me move forward and to consider the right person, place, and time for an answer. And I was even less aware about what I'd been looking for from the astrologer in the first place: not just an answer, but someone, something, to make me feel less alone.

The Information Illusion

At the start of the pandemic, physically isolated from friends and family, fighting daily with A. in our small apartment, and wondering whether I was destined to live out that first astrologer's prediction, I obsessively read articles about relationship repair and communicating with a partner who was different. We had the same fights again and again, mostly about how we were, or weren't, showing that we cared about each other. It was as if we were assigning each other daily performance reviews: *Does not meet expectations*.

To me, showing care meant validating my experiences. To him, care meant critically exploring them. Whenever I was struggling with something at work—say, a micromanaging boss or a toxic colleague—I would share my assessment of the scenario and what I wanted to do about it. A.'s default reaction was not to nod and tell me that this sounded like the right course of action but to interrogate my assessment. This reaction made me feel threatened and defensive: Why didn't he think I knew best what to do in my own life? On the flip side, if A. told me about a situation that was bothering him, I would often reinforce whatever he was already thinking. "But what do *you* think?" he would ask, growing frustrated, expecting me to apply the same intellectual rigor I used in my work to his problems.

We both wanted the other person to accept us for who we were. But we also wanted the other person to change to fit our own subjective expectations of an ideal partner.

Over time, the constant critiques and arguments sucked the marrow from our relationship, making it feel dry and brittle. It also laid bare a paradox: What made our relationship special—the connection we had over inquiry and a willingness to explore uncomfortable questions—was also what made it tumultuous. A.'s tendency to be skeptical and question could turn into confrontation and argumentation—which for me was unfamiliar and scary terrain. In the past, I'd dated people with backgrounds and life experiences nearly identical to mine, which meant we often saw situations in the same way. Not so with A., who almost always had a different perspective. This difference was both interesting and unsettling, forcing me to sit with more uncertainty about our relationship, my decisions, and my own point of view than I ever had before.

So I sought to eliminate the doubts with information from experts, who I hoped would hold answers for what to do. I followed relationship gurus, subscribed to their podcasts and newsletters, highlighted books that promised the secrets to satisfying marriages.[8] Most of all, I was hungry for stories of people in circumstances like mine—living with a partner who made them feel equal parts love and doubt. I started to fantasize about recording our fights, sharing them with a bunch of people, and asking them not only who was right but also whether we should, in fact, still be together. Even with all of this information consumption, I was still stuck, and lonelier than ever.

This feeling was familiar to Mateo—a twenty-something Latino man I connected with through a support group for people with technology addictions. He'd spent years consuming endless amounts of information online to resolve questions the internet simply couldn't answer, unaware of the deeper reasons for his hunger.

Mateo began using the internet when he was a kid to help him cope with the difficulties of making friends. "My relationship with the internet filled this big hole around my struggles connecting with other people, with trusting people around me," he told me.[9]

There, he could be invulnerable. "A huge fear of mine was always that I would look stupid, that I wouldn't know something. On the internet, no one would know I had these basic questions."

Growing up, Mateo read books voraciously, feeling a thrill whenever he learned something new. But the older he got, the more he retreated into digital sources to feed his hunger for knowledge.

"I knew something was wrong about my behavior when I was asking the internet questions I knew no one could answer, like 'Why don't I have any friends?' Or 'How can I be happy?'"

The drive to scour the internet for the answers to unanswerable questions started when Mateo was in college. He was struggling "to connect with other people, to feel authentic or comfortable in my own skin" and was beginning to realize that maybe his childhood had been different from other students'. Mateo was abused as a child and was forced to watch other children be abused, too. During those early years, "I remember feeling like I had fallen into a hole and would never come back out," he recalled. The experience made him desperate to escape reality. "I felt like I didn't want to be me. A part of me felt responsible for what had happened, especially to the other children. I came to see myself as a dangerous and bad person."

His solution: Become someone else, a person who could "completely control my own self, my environment, and other people so it would never happen again." He figured that a person who knew all of the answers, who had a total command of knowledge, was a person who couldn't be hurt or allow others to be.

Sitting alone in his dorm room, Mateo still felt far from knowing all of the answers. Lonely, depressed, and heavy with big questions, Mateo would sift through social media, news, blogs, online research, podcasts, documentaries, and reviews, endlessly scrolling feeds, sensing that an answer was within his grasp. He'd read through hundreds of responses on sites such as Yahoo! Answers and Quora, even the responses that didn't apply to him, desperate for anything to help him feel better.

On the subject of happiness, "people would bring in all kinds of bullshit, random self-help answers, like 'Happiness is moderation,'" he recalled. Though he acknowledged that for some people, happiness *could* be moderation, this notion wasn't relevant to his own particular flavor of unhappiness. "It's impossible to find the answer to happiness on an internet forum. But I would just drink it up and look for more and more of it."

Sometimes he would stay up all night, doomscrolling until the sun rose. Afterward, "I would come out feeling awful, saying 'I'm never going to do this again.' But then, a few weeks or a few hours later, I would start again. I would have moments where I was watching my hands moving on the keyboard and thinking, 'Please stop, please don't do this anymore, I'm tired, I'm in pain, what I'm consuming is meaningless.' I felt like I was trapped behind my own eyes."

Continuing to believe that solutions for his poor mental health could be found with more information, he spent days searching online for ways to alleviate his depression, scouring blogs and forums in a feverish quest for optimization. He purchased dozens of giant vitamin supplement bags and tried other purported cures: cold plunges, new workout routines, light therapy lamps, breathing techniques, tips for how to connect to other people. They didn't work. Around the time of the US election in 2016, he became tethered to the news, constantly refreshing his feeds, reading the same stories across dozens of media sites. "It felt like I was trying to get my hands around this huge thing that was so overwhelming to me," he says.

That huge thing was uncertainty. He saw it as a threat, something to eliminate. To him, appearing uncertain was an undesirable trait, one that someone seeking more friends would do well to shed. "Culturally, people who know a lot tend to be celebrated more than people who don't," he told me. "It's an attractive feature." On the flip side, "being someone who didn't know a lot felt dangerous and ugly."

It also felt profoundly scary. As someone who had experienced abuse as a child, "there was something deep in me that saw the world as an unsafe place, and for me to not be in control of my environment and other people felt potentially life-threatening." Though he knew his quest to eliminate uncertainty from his life through internet searching was impossible, it still felt better than nothing. If he wasn't actively searching, he was "consigning myself to the unknown, which meant getting stuck or being vulnerable."

And there was another illusion that kept him continually coming back to his devices: the sense that he was learning something, or even growing, from his consumption of answers. But amassing facts does not always translate into learning or the kinds of deeper behavioral and emotional shifts he was seeking. Online, the answers he read were robotic, sterile, "divorced from who I was and what was actually going round me."

Decades of evidence indicate that sustained learning and growth stem from real experience and in the context of a trusted human relationship.[10] As he spiraled deeper into internet answers, Mateo faced a cruel paradox: The more he spent his time and energy researching how to connect with other people, the less approachable and more isolated he became. Reading people's questions, answers, and posts made him feel superficially less alone, but it wasn't helping him connect more to people in real life or get out into the world to have experiences of his own. In fact, it was doing the opposite. After being up all night reading, the next day he would show up to his work and other commitments as a zombie, "terrified that someone else would find out what I was actually doing. I would end up feeling more afraid and distant from people in my real life."

He began to consider the idea of a deep, trusting relationship to be a collective fiction, a kind of abstract and idealistic concept fueled by advertising. No one could *really* have friendships like the ones he saw portrayed online, he thought. Surely, everyone else was carrying around the same weighty backpack of loneliness.

Here, then, was his cycle. His fears surrounding his relationships, his commitments, and his responsibilities would catapult him away from the analog reality and into a kind of digital purgatory. Online, he felt he was able to "stop time, to just rest and catch my breath. To disappear from people who might judge me, or my own failures." The longer he spent in these voids, "the more terrifying the world outside would seem, and I would need to escape even more."

The more he tried to find control and certainty through the safety of information online, the less he had. He became actively suicidal, mired in shame about the addiction he kept secret from everyone he knew. Mateo told me he would travel to bridges in the middle of the night, looking over the edge for a while before trudging home. Sometimes when he was cutting something, he'd hold the knife up to his chest. He'd pause, feeling some relief and sadness, before continuing to chop. "I couldn't control myself, and tried everything," he said. "I felt like a huge fucking idiot."

Salt, Fat, Sugar, and Information

Mateo's addiction wasn't just to the internet. It was bigger than that. The addiction was also to information itself—and to the illusion that, if he just spent enough time researching, he could find certainty, comfort, and connection; that with just one more Google search or by reading through one more thread on Reddit or Quora or exploring one more YouTube channel, he would feel better, because he'd obtained answers to his questions.

When we don't know something, there can be an illusory comfort in seeking as much information as possible, one prominent internet addiction psychiatrist told me. The search isn't all bad, especially if you're able to learn from others' experiences or discover support networks. But the trick is recognizing that the internet doesn't have the answers to everything—and particularly to life's biggest questions. The psychiatrist told me that many of the people who he saw were lonely, but used the internet indirectly to address their loneliness. It was, he said, kind of like eating junk food when you were hungry—it might fill you up, but it wouldn't give you long-term nutrition.

The bottom line: The internet can be the right place for some answers but not others. It takes wisdom to know the difference and to understand what you're really seeking from your search.

Though there is no consensus on the terminology, categorization, or diagnosis of internet addiction, one recent study estimates that 2 percent of adults worldwide exhibit behaviors consistent with this condition.[11] But this statistic is hardly representative of how the rest of us *feel* about our reliance on our devices and the subtle ways we find them interfering with our lives. One Pew Research study indicates, for instance, that more than three in ten adults say they're online "almost constantly."[12] That number was 21 percent in 2015. Today, the psychiatrist told me, the internet addict includes all of us.

The internet is addictive in part because of how our brains are wired to respond to new information—which, online, is always just a click or scroll away. In 2019, researchers at UC Berkeley's Haas School of Business found that the discovery of novel information affected the brain's dopamine-reward system the same way as money or food did.[13] "Just as our brains like empty calories from junk food, they can overvalue information that makes us feel good but may not be useful—what some may call idle curiosity," study author Ming Hsu, associate professor at Berkeley, said at the time.[14] Hsu is a neuroeconomist, who studies the intersection of neuroscience and decision-making.

For this study, Hsu and then–graduate student Kenji Kobayashi, now a postdoctoral researcher at the University of Pennsylvania, used functional magnetic imaging (fMRI) to scan the brains of people playing a gambling game. The game presented an opportunity: Participants could learn more about the odds of winning a lottery, but only if they were willing to pay. They had to decide how much they were willing to pay for the information. In some cases, the information was valuable, such as when participants discovered they had a high probability of winning a big pot. In other cases, the information held less value, as when the potential pot of winnings was small. Whether or not it pointed to a large prize or had an impact on their final decision, the information activated the same regions of the brain that are also turned on by food, money, and drugs. The key takeaway: Information is an addictive substance—in part because more information makes us feel more certain about decisions. Sometimes that certainty is warranted: Information *can* help us make wiser decisions. Other times, more information doesn't always help, even though it feels like it does. Information's usefulness depends on its source and whether you've defined what you really need to move forward (for instance, whether you've realized that what you're really seeking isn't just to make a decision but also to feel less lonely).

The consequences of information addiction can be benign: Perhaps you lose a day to watching "how to be happier" videos on TikTok, while some part of you knows that the messages you receive probably won't, in fact, make you happy.* We all do this sometimes, chasing the sense of hope that comes from uncovering a new list of hacks, tips, and tricks that promises to douse a smoldering problem.

Or the consequences can be dramatic and life-threatening, as in Mateo's experience. As we began to see in the first chapter, our quest for fast, easy answers to our deepest questions can make us more vulnerable to cults, conspiracy theories, and misinformation—to the Charlatans of Certainty who claim they have all the answers.

And increasingly, new technologies not only take advantage of our desire for answers but also exploit our hunger for connection and community in our most vulnerable moments.

Follow Me for Answers

Ciera Kirkpatrick always wanted to be a mother. But she was particular about *when*. "I always had this mindset that I would do things in the right order," she told me. "I wanted to be able to achieve certain things so I could provide for [my kids]."[15]

What she wanted to achieve was a PhD. Timing an academic career with children is notoriously challenging, so Kirkpatrick decided to do both at the same time. She knew it would be stressful but figured she could handle it. Once she became pregnant, she began following other mothers on social media—mom influencers—and listening to podcasts. Always the A student, she was determined to do all the right things to give her future son the best life she could. In 2020, Kirkpatrick joined what was an already vibrant community: According to Edison Research, 87 percent of US moms were social media users in 2020.[16] In 2023, that number jumped to 93 percent, compared to 70 percent of the general population.[17]

---

* A quick perusal of this content reveals these keys to happiness: journaling, gratitude, going to sleep early, eating nutritious food, and exercising. Not bad advice, but probably not what's needed to address deeper questions or mental health challenges.

Much of the content she consumed before giving birth was focused on breastfeeding, "the importance and benefits, the beautiful journey you and your child get to go on together," she recalled. The message from all this media was clear: There was a right way and a wrong way to nourish her child.

Her son was born in May 2020, at the height of the pandemic. After taking him home, Kirkpatrick and her husband noticed he was having trouble breathing, making it hard for him to latch to Kirkpatrick's breast. "I felt like I was as prepared as I could be," she said. "But you start to realize [early on] how things don't go according to plan. The things you believe are going to console your child or make sure they are fed don't work."

Less than twenty-four hours after being sent home, a visit to her son's pediatrician resulted in him being transported by ambulance back to the hospital because he had lost too much weight. After her son spent another five days in the hospital, and countless tests, the doctors determined he was suffering from extreme dehydration." We were pushing breastfeeding so hard, but he was starving, and wasn't able to get the nutrition he needed that way," Kirkpatrick said. "He almost died. You feel so much pressure to do things the right way, but the right way isn't necessarily the right way for you and your child. So much of handling the new role of being a mother must come from your experience."

For Kirkpatrick, now an assistant professor of advertising and public relations at the University of Nebraska–Lincoln, who studies how media messages influence individuals' health outcomes, the experience was a wake-up call about the pervasive influence of social media on her decisions as a mother. This became the focus of her PhD dissertation and subsequent research. In a 2022 study, she exposed 464 new mothers (with a child three years old or younger) to twenty Instagram posts from what she labeled "mommy influencers" and "everyday mothers."[18] Half of the posts presented idealized depictions of motherhood, with glossy, made-up mothers, grinning children, and spotless homes. The other half were more realistic portrayals. The mothers exposed to the idealized posts—whether they were created by "mommy influencers" or "everyday mothers"—experienced increased anxiety and envy.

Other research suggests that while mothers are cognizant of wanting to trust only reliable websites for health information, they didn't express the same concerns about social media.[19] Maybe because the content in those Facebook groups, blogs, and listservs was written by people whom mothers perceived "to be *just like me*, mothers in our study considered the information and opinions expressed on social media as being trustworthy, perhaps even more trustworthy than those of health care professionals," write the researchers.

This urge and tendency to trust social media moms is understandable, Jessica Grose told me.[20] Grose is a *New York Times* opinion writer and author of *Screaming on the Inside*.[21] For one thing, the sleep deprivation alone is enough to warp anyone's ability to make decisions and distinguish fact from fiction. "In actual cults, they don't let you sleep so you'll become more suggestible," Grose pointed out. Another factor is social isolation, which was heightened during the pandemic but may still be a challenge for new mothers who live far from family or friends, or who don't have people in their lives whom they trust for parenting advice.

Motherhood, Kirkpatrick told me, "can be very isolating at times" even if you *do* live near family and community. "You're likely experiencing a major change to your routine and social life. For instance, you might not be going out with friends and co-workers like you used to, and you might not be going to work like you used to. Just being home and alone with a child all day can be isolating in itself—just because of the changes to your routine. You can feel really alone in your emotional state and the struggles you're going through."

In the context of isolation or dearth of community support, women turn to "experts." In her book *Making Motherhood Work*, sociologist Caitlyn Collins found, in a qualitative study of 135 working-class mothers in Sweden, Germany, Italy, and the United States, that American mothers are more likely to reference expert perspectives when talking about the definition of a good mother, rather than presenting their own.[22] The bulk of Collins's study explored how different social policy contexts influence the way women deal with work-family conflict in their lives, unveiling distinctive desires, expectations, and norms in each culture. When asked what being a good mother meant to them, many American women cited books, podcasts, and articles, rarely sharing their own definition.[23] These responses were in stark contrast to those of European women, who instead talked about the character traits they hoped to cultivate in their kids.

The point here is not to say that experts are wrong or shouldn't be trusted. It's about the importance of considering whether or not an expert is the *right one* to answer *particular questions*. It's about becoming less reactive and more responsive in our uncertainty, so that we're able to better understand what we're really looking for when we seek an answer and whether we're likely to find it where we're looking.

Since having her first child, Kirkpatrick's relationship to social media and mom influencers has shifted dramatically. In the years since her son was born, he was diagnosed with autism. At age four, he is still largely nonverbal. Kirkpatrick has sought support from online communities of parents of children with special needs and influencers with children like her son. But today, she's able to better discern the answers online communities can give her from those that she needs to discover inside her own experience. Especially because, as she notes, no two children with autism are alike.

"In our situation, we have no idea if our child will ever communicate verbally," she said. "We have no idea what the future will look like for him. Will he have a job or a relationship? Will he live with us for the entirety of his life? What happens to him if something happens to us? Is this something my marriage will survive?" Of course, these are pawpaw-style questions that will take time to answer and understand. But in the meantime, Kirkpatrick said, she's found that having the right support and community around her—online and offline—can "help a lot in dealing with the uncertainty." Bolstered by those relationships, she can take the time she needs to sit with her questions, knowing that others have eventually found answers to those questions, too.

Want to Chat?

The Charlatans of Certainty are masters of adaptation. They've built social media influencer empires just as they captured attention in the early days of the internet, ready to feed information addictions with their promises of answers. This evolving behavior of the Charlatans raises a question: What could these players look like in the future, and how do we prepare to deal with them in the present?

Let's first imagine a man years from now, coming home from a long day at a job that feels empty and directionless. He is lonely, having just gotten divorced. And he sees an advertisement during his subway commute. "Let me ask you a simple question," the voice in the ad says. "Who are you? What can you be? Where are you going? What's out there? What are the possibilities?" The ad is for a new artificially intelligent operating system designed to help its owner answer those big questions. Later, the man installs the new operating system and begins to form a relationship more intimate than even his closest human relationships.

This is the plot of the 2013 movie *Her*.[24] No longer pure science fiction, there is a clear path from the world of social media influencers and conversational AI into that world, where we form bonds with AI systems and devices that know all of our preferences, can anticipate our needs, and will happily answer any question we ask, morphing from personal assistant to therapist to significant other in seconds.

Our relationship to AI is poised to supercharge the addiction we already have to seeking certainty through information. By giving us a version of connection, AI may also continue to obfuscate what we're really looking for when we seek that information: real, human relationships.

Long before *Her*, the future the film painted was one that Joseph

Weizenbaum predicted and worried about back in 1966. Back then, Weizenbaum may have felt he created a monster. Something he'd spent months slaving over was working—too well.

Weizenbaum had created the world's first chatbot. The Massachusetts Institute of Technology professor named it ELIZA, after Eliza Doolittle, the main character of George Bernard Shaw's play *Pygmalion*, which later inspired the musical *My Fair Lady*.[25] "The idea was that my language analysis program would keep getting better, more refined, more exact, more varied, just like the flower salesgirl in the musical, under the tutelage of her teacher Professor Higgins," Weizenbaum told Gunna Wendt in *Islands in the Cyberstream*, an extended interview with him originally published in German in 2006.[26]

Weizenbaum programmed ELIZA to act as a Rogerian therapist. For him, this was a form of parody; he believed that this therapy style was robotic and that if he programmed a chatbot to perform it well, others would come to share his perspective.[27]

Rogerian therapy, named after its progenitor, psychotherapist Carl Rogers, is characterized by subtle and compassionate guidance.[28] The therapist often reflects what the patient says, asks them questions, but does not judge them or offer prescriptive advice. The big idea in Rogerian therapy is that everyone has the tools within themselves to grow and change, and the therapist serves as a trusted guide, rather than an omniscient authority. ELIZA extracted key words from its conversation partners to generate questions, an essential component of Rogers's concept of active listening.

The result, for Weizenbaum, was "astounding." People *loved* ELIZA. Despite Weizenbaum's protestations that ELIZA was *not* human, people were still asking to speak with it privately, recounting moments of therapeutic profundity.[29]

"The most extreme experience was with my secretary," he recounted to Wendt. "Once I came into her room and she was in the middle of a 'conversation' with 'Doctor.' She reacted in a way I could not understand at all. It was visibly obvious that I was interrupting her session and after only a short time she asked me to leave her alone. It was as if

I had disturbed an intimate moment. I found it absurd because she had witnessed the development of this program up close. Hardly anyone—except for myself—knew better than she that it wasn't anything more than a computer program."[30]

Deeply disturbed by how quickly people had connected with his machine, he shut down the program and became a prominent critic of artificial intelligence, later writing unequivocally of the dangers of trusting computers with our deepest questions: "We have permitted technological metaphors . . . and technique itself to so thoroughly pervade our thought processes that we have finally abdicated to technology the very duty to formulate questions," he wrote.[31]

As a German-Jewish immigrant who had escaped Nazi Germany with his family in 1936, Weizenbaum felt it was his duty to speak out when he saw the potential for science to create societal harm—to never be like the German scientists who shirked responsibility for their role in the atrocities of the Holocaust.[32]

He was not antitechnology but rather a proponent of considering not just what technology *could* do but also what it *should* do. He questioned whether the advancement of our tools was accelerating our psychological and spiritual progression as a species.

"No other organism, and certainly no computer, can be made to confront genuine human problems in human terms," he wrote in his 1976 book *Computer Power and Human Reason: From Judgment to Calculation*, perhaps anticipating the kinds of questions that AI would lay claim to in *Her*: "Who are you?" "What can you be?"[33] In a 1985 interview with the *New Age Journal*, he went even further to shame the computer evangelists.[34] "The dependence on computers is merely the most recent—and the most extreme—example of how man relies on technology in order to escape the burden of acting as an independent agent," Weizenbaum said in the interview, years before personal computers became fixtures in every home. "It helps him avoid the task of giving meaning to his life, of deciding and pursuing what is truly valuable."

His words echo in warnings from former Secretary of State Henry

Kissinger, former Google CEO Eric Schmidt, and MIT Professor Daniel Huttenlocher, who nearly sixty years after ELIZA wrote about conversational AI in *The Wall Street Journal*.[35] Collectively, we need "to reflect on which questions must be reserved for human thought and which may be risked on automated systems," they wrote. "We must be thoughtful in what we ask of [conversational AI].... Humans will have to learn new restraint."

There are trade-offs when we always turn to our devices rather than other humans to answer questions. When we ask a question—particularly one without an objective, correct response (think again about our questions tree and the heartwood-style questions, or even the pawpaw questions that may take longer to ripen)—answers are only part of what we're seeking. Often, whether we're aware of it or not, we're also seeking to connect with others. Recall Mateo's loneliness in the midst of his information seeking or the isolation of new mothers who turn to influencers for their queries.

To Xuan Zhao, research scientist at Stanford SPARQ (Social Psychological Answers to Real-World Questions), overreliance on AI to answer questions could have negative consequences for human connection. Her research suggests that we underestimate people's willingness to help or give us advice, and that these interactions can be deeply meaningful for both the asker and the helper.[36]

"When you ask questions of other people, often times the conversation itself is an opportunity to connect," she said.[37] When you turn to conversational AI, "you may risk losing a chance to connect with other people."

To address this risk, Zhao decided to cofound Flourish Science, a public benefit corporation that makes facilitating human connections one of its core principles. "If I could wave a magic wand, I would want to live in a world without AI or even smart phones," Zhao said. "But that's not the world we live in. So as a social psychologist, I feel it is my responsibility to think about how to better harness AI before it is turned into products that exacerbate social isolation." Zhao and her team created Sunnie, a conversational AI in the Flourish app, "to inspire people to better connect with themselves, others, and the world."

The trick is not to try to avoid emergent technology entirely, which, as Zhao suggests, is not possible for most of us. Rather, our task is to more deeply consider which questions we choose to ask of it. Just as I learned from Lanyadoo, the astrologer, we need to ask *ourselves*, "Am I asking the right question of the right source?"[38]

A life spent seeking certainty and chasing after answers can have dramatic and sometimes painful costs. In conversations with Mateo, Kirkpatrick, and others, I kept hearing a central message: What makes us vulnerable is not just that we tend to seek certainty and avoid uncertainty, but also that answers to our questions are only part of what we want or need. Sure, we're often looking for information. But we're also motivated by other drives—such as our desires for connection, belonging, community, and meaning. Lacking awareness of these deeper motivations when we search for answers can make us more vulnerable to exploitation by people and communities that claim they are able to provide answers and everything else we need to thrive.

"A big part of my recovery has been not just looking at surface behaviors, like internet binges, but at the deeper needs I'm trying to meet through these behaviors," Mateo told me, emphasizing that his membership in the technology addiction support group is playing a core role in the healing process. "Being away from all those behaviors, my fear was that my life would get small, that it would get boring, banal, mediocre, and disconnected. But the opposite of that happened."[39]

Today, instead of watching the lives and experiences of other people, "I get to live my life, and it feels exciting, challenging, meaningful, and difficult. I have deep relationships with other people. I don't feel like I have to navigate a spiderweb of lies."

Before his recovery, he had friends and girlfriends. But these relationships were limited, hollow, because he'd never been comfortable being what he felt was a truer version of himself around these people: someone who didn't know all of the answers, who sometimes felt uncertain, sad, and afraid. Instead, "I was an intimidating, mysterious, and withdrawn person to a lot of people," he said. He was particularly drawn to other addicts or narcissists, people "who weren't able to give me what I needed but promised a sense of deep connection, which I longed for but didn't know how to have."

Mateo's answer-seeking behavior was an attempt to shield himself from others' perceptions that he might be uninformed, ignorant, somehow less than. To him, being *right* became connected to being accepted and loved—to a sense of ultimate security.

But now, in recovery, "I have unlimited access to people telling me not that I'm right, but that I'm *all right*. I'm okay as I am. I don't need to hide my doubt and my insecurity and my lack of knowledge. I was so obsessed with knowing the right answer to get people to like me, but it's actually so much more pleasant to talk to someone who knows that they don't know the right answer." He mentors other people in his recovery program. His life, he told me, is filled with connection—not the shallow kind he used to have but deep friendships with people he knows he can call in a crisis.

And there's another relationship he has transformed: his relationship with himself. Even several years into recovery, the furthest he was able to get was simply not hating himself or actively wanting to die. But today, he said, "I really love myself."

Then, he pauses. "Honestly, saying those words would have made me vomit in the past." I know what he means—talking about loving yourself can feel contrived and corny. For him, the feeling is real, and it has changed his life. He credits the recovery community for helping him accept the parts of himself that he was most ashamed of and helping him see that he would never be able to find the deep relationships he craved if he couldn't forge that kind of relationship with himself first.

Loving himself, he discovered, was also intimately tied to his relationship to uncertainty. "When I'm at home in my own body and own life, I don't need to know more than I know, or be more than I am," he said. Part of what used to drive his desire to know the answers was a need to know what would happen next and fears "around losing something I have, or not getting something I want. When I can just accept reality and myself as I am, I can feel more at peace in uncertainty."

With this changed relationship to himself, Mateo has also realized that he doesn't always need to search for answers outside—that "most answers come from within."

It's worth pausing for a moment on this last point. It might sound like an obvious, self-help platitude—the kind that would have made Mateo puke in the past or that he would desperately consume in his online searches. But it's not always clear in a culture that encourages us to seek knowledge elsewhere, that bombards us with constantly flowing content feeds and advertising designed to make us doubt our worth and decisions.

It's also not always clear just *how* to find those answers buried deep within. Or at least it wasn't for me—until I began to meet a group of people who used their questions as a kind of internal GPS.

PART TWO

LOVE THE QUESTIONS THEMSELVES ➤

THE REWARDS OF COMMITTING TO CURIOSITY

4

What Does a Life of Loving Questions Look Like?

THE BIG IDEA

The relationship you have with uncertainty is a proxy for the relationship you have with yourself. The act of loving your questions represents a commitment to listening to and learning from yourself that leads to greater self-compassion and self-integrity.

Rilke has been ill for weeks. At the beginning of July 1903, he decides to leave Paris, hoping that the fresh air in Worpswede will heal him. There, in Worpswede, he sits down to write his fourth letter to Kappus.[1]

"Rilke was hardly qualified to give career advice at that point in his life," author Rachel Corbett writes in her book *You Must Change Your Life*.[2] He had turned in his monograph on Rodin the previous December, but he still "could not even afford to send friends copies of his books." And yet, Kappus's letters touched something in Rilke, perhaps because he, too, had been feeling lost, isolated, and uncertain in Rodin's Paris, struggling to understand who he was beside the great master, someone Rilke thought would give him all of the answers.

Though Rilke, at twenty-seven, is still "sublimely naive," he responds to Kappus "in a tone of authority that only an amateur would dare—trying on the master's robe and liking the way it fit," Corbett writes.[3]

"You are so young, so before all beginning," he writes to Kappus, before instructing him, famously, to "be patient toward all that is unsolved in your heart and try to love the questions themselves like locked rooms and like books that are written in a very foreign tongue. Do not now seek the answers, which cannot be given to you because you would not be able to live them. And the point is, to live everything. Live the questions now. Perhaps you will then gradually, without noticing it, live along some distant day into the answer."⁴

Corbett believes part of this philosophical strategy—to live the questions now—was inspired by watching Rodin's approach to sculpture, "this idea of taking a hunk of clay, or a block of marble, and not knowing what he's going to make." Rodin would "chip away until a face was formed, or a leaf. Rilke described it as 'a worm working its way from point to point in the dark.' Rodin liked to leave his sculptures half formed because he loved the process of the question—the wondering and then the becoming."⁵

It was the summer of 2014, and I felt like I'd just stepped into a nightmarish parallel universe. Nearly twenty-six, I'd just broken up with a serious boyfriend I had been dating for two years. I'd agonized over the decision for months. I still cared about him, maybe even still loved him. And that's what had been so confusing: On paper, he was the perfect guy. Part of me thought we would get married; at one point, he told me that he would get down on one knee whenever I was ready. For two years, he had consistently nourished our relationship with date nights, international vacations, sweet gestures, and romantic notes. He had a great job, a great family. He was smart. We'd both been raised Jewish in similar socioeconomic environments. It *should* have worked. I wanted it to work! My brain was telling me to *make it work*. But my body had a different message: Something wasn't right. At the time, I had trouble putting my finger on exactly what wasn't working about our relationship. Now, I think I understand better: I needed a partner like A., who would challenge me to grow, both intellectually and emotionally. My ex rarely asked me any questions and seemed to have decided early on who he thought I was, putting that idealized version of me onto a pedestal. To me, our relationship felt too comfortable.

After we broke up and I plunged into the uncertainty of single life, I (of course) missed the security of the relationship and missed him. The whole thing felt complicated. Some moments I knew that what I'd done was the right thing; other moments I doubted it. I relayed some of these emotions to my friends at the time, including one of my closest friends, L., the one who had introduced us. We were *all* a tight group, so this breakup wasn't easy for anyone.

A few months after we'd officially ended things, L. texted me that we needed to talk about something. Usually this meant that she needed some dating advice. I texted back, telling her I'd love the company: The day before, I'd tripped on some uneven sidewalk outside my studio apartment and badly sliced open my knee, making me somewhat housebound.* Hobbling to the door to let her in, I noticed that she seemed a little nervous as she sat down. I plopped on the couch across from her, elevating my bandaged knee.

Breathing in and out deeply as if she were in the middle of a strenuous yoga pose, not meeting my gaze, she told me that it had all started on the Fourth of July, after the breakup. The identity of "it," at first, was vague and ominous. She spent the holiday with my ex-boyfriend—who remained her close friend. But that night, the relationship had changed. They'd kissed. And since then, for the past month or so, they'd been hooking up. Now, she told me, they were going to be together, officially. She wanted me to know how much she still cared about me, how hard it was for her to do this. But in her heart, she knew she had to give this relationship a chance.

---

* At this point in the chronicling of my injuries and accidents, some of you may be wondering whether something is wrong with me. I've wondered the same thing. In my mid-twenties, I tripped and fell on several runs over the course of a few weeks and went to a doctor to see whether I had some kind of nerve or balance issue. After checking me out, the doctor told me that the only diagnosis he could offer was that I was simply a bit of a klutz. He didn't use those exact words, but you get the gist.

I felt a set of paradoxical reactions mingling inside of me: both shock and the sense that this all felt very familiar. I had long suspected something was going on between them; I'd even asked him about it once. She and I happened to look a lot alike, and there was often an energy between them that made me a little . . . suspicious.

Now here she was, confessing. And yet, could I really be angry? I was the one who had broken up with him, after all. Still, among many women, there exists an unspoken rule: you don't date your close friends' ex-partners, particularly when the breakup is still fresh and *especially* when you know how complicated the circumstances were around the relationship.

As my compulsively agreeable nature kicked in, I managed to say something about how, sure, I would support them and hoped they would be happy, along with some other platitudes that didn't really feel right but that seemed like the things I *should* say.

After she left, I felt more alone than ever. And then, the numbness settled in. I didn't cry. I didn't rage. I told a few friends what had happened, vented in a kind of detached way, and tried to get on with my day, hoping I could just forget about it.

When we think of betrayal, we tend to think of stories like this: a friend or partner who acts to break our trust. But what made my situation hard to process was not just the capital-*B* betrayal of my friend and my ex. It was what happened afterward. It turns out that what happened afterward was mostly a continuation of what had happened before. Every day, in ways big and small, I had been betraying myself, living a life divided between who I thought I should be, what I thought I should want, and who I really was.

And I was not alone in this pattern, particularly among people who have learned—often through painful experiences—how to love their questions through a long ripening process. When I faced heartwood-style questions about myself, ones that felt intimidating and impossible to answer, I didn't sit with them lovingly and curiously, open to their presence in my life. Who had time for that? To answer them, I turned outward instead of inward. Instead of "What do I want?" I asked, "What *should* I want?" Ideas of who I *should* be—whom I *should* be dating, where I *should* work, how I *should* act—guided my decisions. They were forms of fast, easy answers.

I received external validation whenever I made one of those "should" choices, so they were hard to turn away from—especially because I wasn't sure what I was turning toward, exactly.

This, too, was Parker Palmer's dilemma.

Out of the Darkness

As a young man, Parker Palmer was a golden boy: the student council president who was accepted to a top college and graduate school, always graduating with honors. He attracted powerful mentors—foundation executives, college presidents and deans, people who saw a future educational leader in him, a younger version of themselves, someone to mold.

For a while, he'd followed the path expected of him—earning a PhD in sociology from Berkeley in preparation for an academic career. But by the time he received his degree in 1969, US cities were burning from racial injustice, and Palmer felt deeply called to use his sociology in the streets rather than the classroom. So he veered off course: he became a community organizer in Washington, DC, pushing back on redlining and blockbusting with strategies for building community.

During this time, a big question began hovering around Palmer's life. "The people who knew me best, including my friends and mentors," Palmer said, "began finding ways to ask, 'What on earth are you doing with your life?' Despite the depth of my calling, the question haunted me, because for so many years I didn't have an answer that made sense to others. All I could say was, 'I can barely explain this to myself, so I'm sure I can't explain it to you. I simply have a deep sense that I can't NOT do what I'm doing.'"[6]

At stake in the question were his identity and purpose. Who was he, anyway? What was he supposed to be doing? These formative questions haunted Palmer, just as they did Franz Kappus, Rilke's nineteen-year-old correspondent.

For five years, Palmer worked in DC as a community organizer, until he began feeling burned out. In 1974, at age thirty-five and intent on learning more about the inner journey and the practice of nonviolence, he took a one-year sabbatical with his wife and three children at a Quaker living-learning community near Philadelphia called Pendle Hill.

Though Palmer and his family thrived at Pendle Hill, the question that vexed him—*What on earth are you doing with your life?*—only intensified. When his sabbatical came to an end, the question remained unanswered, but, happy with their communal lifestyle, he and his family decided to stay. In 1975, Palmer was hired as Pendle Hill's dean of studies. He became responsible for managing a teaching staff of about six people and a curriculum for resident adult learners. They could take classes on topics ranging from meditation, prayer, and poetry to the history of world religions and nonviolent social change, all in the context of a full round of daily communal life.

"A lot of the adult students were there at critical junctures in their lives," Palmer explained. "Maybe someone had been burned in a career and was looking for a new direction, or had recently been separated or divorced, maybe a young person who didn't know where to go from here." Each of them, like Palmer, was in the midst of a transition, holding big questions for which they did not yet have answers. Palmer knew he didn't want to go back to traditional academia, but he didn't know what he would do. He was supposed to be supporting his family of five, who lived on campus with him, but given Pendle Hill's radically egalitarian staff salary structure, he and his wife were barely making enough to make ends meet. Would he be able to support his three young kids the way his parents had supported him?

As those questions grew deeper and good answers eluded him, he began a slow slide into depression. On his worst days, it sometimes struck him that it might be best for everyone if he weren't around, carrying the feelings of doubt and shame that stalked him.

Several years into his tenure at Pendle Hill, he woke up one night from a nightmare he couldn't shake. In his dream, it was World War II, and he was hiding from two Nazi soldiers in an isolated farmhouse in France. The soldiers, flanked by two German shepherds, were trying to kill him. The first Nazi found him cowering in a room behind a piece of furniture. As he and the dog lunged for him, Palmer reached into his pocket and pulled out a little gold penknife, stabbing the dog and the soldier, who both fell dead. Then, the second soldier and dog approached. Palmer killed them, too.

A week or two after the dream, the psychoanalyst Robert Johnson visited Pendle Hill to give a lecture. Johnson was a Jungian analyst, known for his writings and interpretations of the Swiss psychoanalyst Carl Jung. This meant he was particularly attuned to the symbols and possible meanings of our dreams.

He was also, in Palmer's recollection, "a tough and crusty old guy, the kind of man I did not want to be." But this didn't stop Palmer from approaching Johnson and asking him if he would listen to Palmer's dream and see what he thought about it.

"He was not a user-friendly guy," Palmer says. "But I didn't care. I wanted to understand what was going on with me." Johnson agreed to speak with Palmer before his planned lecture. They met in a small apartment where guest lecturers stayed on the campus, and Palmer recounted his dream.

"Tell me about the knife," Johnson said.

"It was a penknife, with a gold handle, and it flipped out—just a simple thing," Palmer told him.

"Have you ever seen that penknife before?" Johnson asked.

Palmer suddenly realized that he had. "My grandpa Palmer had a penknife like that," he said.

"Tell me about Grandpa Palmer," Johnson probed.

"He was a simple man without much education who always told the truth," Palmer said.

Johnson looked at Palmer, and slowly the meaning of what he'd just said sunk in.

"I got the message," Palmer tells me, recalling this moment. "I could save my life from the darkness that was haunting me by telling the truth—telling myself and some of the people in my life the truth about what was going on inside of me. So I started opening up about my fears, and for the first time in my life, I got into therapy with another Jungian analyst."

For Parker, this interaction with Johnson was a moment of insight when he felt called to "get real," to explore, acknowledge, and reveal still-hidden parts of himself. At that time, the "truth" he felt compelled to tell was complex and layered—including struggles with relationships and financial security—but at bottom it was about vocation, his life's meaning and purpose.

Even today, at age eighty-five, the process of coming to terms with his own truth is ongoing, Palmer says. But over his long and fruitful career, he has reaped the benefits of facing it and sharing it with others, along with his experience of living questions that have often led him off the main road and into the woods. After eleven years at Pendle Hill, the Palmers left, and he began to work independently as a writer, speaker, and workshop leader. "Writing," he says, "became a way in which I could continue to tell my own truth, pursuing the clues I first picked up in that dream."

As the years went by, Palmer became an award-winning author and activist—he wrote ten books that have sold millions of copies, founded the nonprofit Center for Courage & Renewal, and received fourteen honorary doctorates and numerous other awards, including the William Rainey Harper Award, among whose previous recipients are Elie Wiesel, Paulo and Elza Freire, and Margaret Mead.[7]

"I'm sure that my work is better known *because* I took the road less taken than it would have been had I stayed on a conventional academic track," he says. "I'm also sure that trying to tell my own truth has brought me more readers than a more detached academic approach would have attracted. And given who I am, I'm certain that walking the road less taken has given me more creative freedom and brought more personal satisfaction than an academic career would have provided."

But though he knew he was doing the work he couldn't *not* do, walking that road—living those questions—wasn't easy.

Facing Our Splintered Selves

Telling ourselves the truth about what we're really feeling, as Palmer did, means confronting where, when, and how we've been lying to ourselves—avoiding answers to questions we don't want to face. The lying isn't always conscious, and it's certainly not always malevolent. Often, it's rooted in a desire to *protect* ourselves, even as the lying creates suffering.

This is one of the conclusions of psychologist Richard C. Schwartz, the creator of Internal Family Systems (IFS), a therapeutic model based on the idea that people are more like a cast of characters in a play rather than a solo act, with multiple parts of our psyche that can be wounded from the past. According to IFS, healing requires us to connect, understand, and integrate all these parts of ourselves. The goal is for our behavior to be governed by our core Self, "an essence of calm, clarity, compassion, and connectedness," Schwartz writes in his book *No Bad Parts*.[8]

What stops us from existing as that higher-order Self all the time? Some of our parts, Schwartz writes, are "protector" parts, which may have developed during a traumatic life experience and can be triggered by situations the protectors think they need to manage because they don't trust your Self to do it. For instance, if you were abused by a caregiver as a kid and unable to stop the abuse, "your parts lost trust in your Self's ability to protect the system, and instead came to believe they have to do it."[9]

Sometimes, these parts hijack your behavior by "blending" and convincing you that you *are* them, not that they are simply a part of you. One part might be an inner critic telling you your latest creative project is hot garbage, a drill sergeant pushing you to work harder even when you're exhausted, a compulsive caretaker demanding that you put the needs of others over your own, or a kind of detached observer disassociating from perceived threats in your day-to-day life. "These symptoms and patterns are the activities of young, stressed-out parts that are often frozen in time during earlier traumas and believe you are still quite young and powerless," Schwartz writes. "They often believe they must blend the way they do or something dreadful will happen (often, that you will die)."[10]

Much of the time, he says, we're not even aware of the influence these parts exert, or how "blended" they've become with our self. Instead, we simply carry around a general sense of, for instance, being unworthy without our accomplishments or unlovable unless we're perfect. "We may not even be consciously aware of such beliefs—yet those burdens govern our lives and are never examined or questioned," he writes.[11]

Questioning the Protectors

I didn't know it at the time, but in my mid-twenties, these protector parts were running my life. This meant I was consistently denying, dismissing, and suppressing my own needs and wants. I had sex when I didn't really want to, with guys I wasn't that interested in. I said yes to happy hours with creepy bosses because I didn't think I had a choice. I said yes to work assignments I knew I didn't have time for. I laughed at jokes that weren't funny. I spent time with "friends" whose company I didn't enjoy. I spoke when I didn't have anything to say but felt I needed to fill the silence. I was agreeable and perky even when I felt like screaming or scowling. In the words of one of my therapists, I was "shoulding" all over myself. In prioritizing the comfort of others, the way I had in the conversation with my friend who started dating my ex-boyfriend, I was growing more and more distant from myself.

If I've learned anything from my own experience and from interviewing others, it's that *a life spent in relationship with questions is a way out of this pattern.* That's because a life of living and loving our questions is less about chasing answers, safety, and protection, and more about pursuing integrity, challenge, and growth. At its core, integrity simply means wholeness. I feel I have integrity, that I am intact, when my actions line up with my values—when I act according to what I want to do, not what I think others expect me to do, and when I'm listening to and caring for *all* the parts of me. For me, this meant that I eventually decided I didn't want to see, or talk to, L. or my exboyfriend anymore. As someone accustomed to putting the perceived needs of other people over my own, this decision felt harsh and even immature. Couldn't I just get over it? But I couldn't—at least not yet. And I no longer wanted to pretend that everything was okay when it wasn't.

Why We Become Divided

It sounds simple, but for many of us, including me, it's all too easy to let doubts that arise from our questions cause us to feel divided between what we want and what we think we *should* want. Perhaps you've felt the kind of queasiness or anxiety of saying one thing but thinking something else, of making a choice that just didn't *feel* right in your body. It feels awful.

We fracture for all sorts of reasons, but author and coach Martha Beck highlights two in her book *The Way of Integrity*: trauma and socialization.[12] Remember how trauma can contribute to our inability to sit with and tolerate uncertainty? It also plays a role in our ability to make choices that are aligned with how we really feel and what we really want.

For instance, when my mom got sick, the lesson toddler Elizabeth internalized was that my mom had left me because I wasn't lovable or because I had done something wrong. How to solve this problem? Of course, the answer was to be perfect! To be the sweetest, best, most agreeable person I could be, so no one would ever leave me again.

Over time, being a people pleaser became a central part of my identity. The more I practiced being a people pleaser, the harder it became to ever imagine letting it go and the less aware I became of its influence on me. I no longer saw the choices in my life clearly. I could see only one way to interact and behave with others, and it became reflexive. I was not openly exploring questions; I was attempting to live out a set of answers about what it meant to be happy, satisfied, fulfilled.

The result? I felt empty, lost, and disconnected from myself and the people around me—even, and especially, as I looked to others to help me figure out who I was or blamed them for who I was not.

In Rilke's words, I was committing an all-too-common mistake. There are moments, he wrote in a letter to a friend in 1904, when it is hard to endure life "within one's own I. It happens that precisely when one ought to hold on to oneself more tightly and—one would almost have to say—more obstinately than ever, one attaches oneself to something external, and that during important events one shifts one's proper center out of oneself into something alien, into another human being."[13]

This sounded familiar. During some of my most uncertain moments, I'd been relying on *other people* to give me answers, to give me stability and security, rather than cultivating security within myself. According to Rilke, "this is against the most basic principles of equilibrium and can lead to nothing but great difficulty."[14]

Palmer had written book after book about balancing community and solitude. I needed to visit him to see how he'd found the equilibrium Rilke had described in his own life, outside of his books. And that was how I found myself traveling to Madison, Wisconsin, on an uncharacteristically warm May afternoon.

You Do Not Complete Me

Though we were born fifty years apart, Parker Palmer and I shared a common origin story. We grew up a few miles from each other and even went to the same high school. And now, we were sitting on his back porch, surrounded by maple and oak trees, reminiscing about those days.[15]

In high school, Palmer recalled, there wasn't much that interested him academically. He mostly earned Cs. But he excelled at extracurriculars: He was the class president every year as well as president of the student council. He was the kid who could deliver a speech to two thousand high schoolers and bring them to their feet cheering.

From high school on, "I got a lot of [positive] projections about who I was," Palmer told me. "And I absorbed a lot of those projections because the heroic projections are pretty nifty. Who wouldn't want to embrace that image of oneself?" When people project, they're attributing their own expectations to you. And in doing so, they're giving you their answers to questions that should be yours alone to answer: *Who are you? Who do you want to become?*

Like finding a fast, easy answer, embracing a positive projection can feel good for a while. After all, it can temporarily resolve uncertainty. But later in life, as Palmer struggled with depression and existential questions about who he was, the life he was leading, and how he could support the people he loved, he began to realize the projections weren't him. They were distortions. When the person he was and wanted to be didn't match up with the heroic projections, he felt guilt, shame, and fear. If he wasn't the hero, what did that mean?

Palmer was caught in a common trap: imagining that he had to be either the hero or the villain,[16] an inspiring leader or an impending failure who couldn't hold his family together. The truth was that he was both shadow and light, as we all are. But this was a truth he was only beginning to discover. Back then, most of what he felt was a sense that he was coming apart.

"The divided life is one where you cut off certain realities from yourself," he told me. "You wall them off from other people, and you're probably walling them off from yourself, because you haven't taken the opportunity to integrate the shadow into the light."

For him, this meant acknowledging that he was someone with gifts and strengths, as well as someone who could fall into a black hole and hurt other people unintentionally. The point, he realized, was not to overemphasize either end of the equation, but to hold it all together so you can "show up first to yourself, and then to other people as, this is me, I am all of the above," he said. "Somehow," he went on, "acknowledging the presence of your shadow gives it less power over you."

For Palmer, wrestling out loud with honest, open questions became a critical tool on his journey toward psychological well-being. "There's no way for me to overemphasize the importance of questions in coming to the place where I am right now," he told me—a place where he has integrated these multiple sides or parts of himself. These were often questions that came up in Jungian therapy—about his own reactions and responses to life, his relationships with other people, and the content of his dreams. Questions that had to do with the meaning of certain symbols and images, or underlying hopes and fears, or emotional responses that went unexpressed in his waking life. "I ended up living some of those questions," Palmer said. "They aren't questions you can answer the way you answer a math problem. They are questions you wrap your life around, questions that increase your awareness of how you are living and why. This kind of question is a constant probe into your experience, and a way of stirring the pot to see what else is in there."

Palmer was no stranger to the power of questions. He first read Rilke's *Letters to a Young Poet* in his early thirties, around the time that he traveled to Pendle Hill. It would take Palmer many years to begin to love the questions this experience provoked in him, in particular the one that caused him so much anxiety: *What on earth are you doing with your life?*

I asked him about that question, and he told me that today, in his mid-eighties, he has an answer that works for him: He's committed to defying the death-dealing cultural trends he sees and to working for "life-giving causes." His books chronicle the norms he has pushed against as he has worked for social and racial justice across his career. In *The Courage to Teach*, he encourages educators to consider not just what and how to teach, but who is doing the teaching.[17] Who is the person from whom teaching comes? Good teaching, Palmer writes, requires deep self-knowledge. In *Let Your Life Speak*, he writes about the power of living "divided no more," of not letting yourself be guided by external pressures to be someone you're not.[18] "If you're in the world with deep respect for self, and you're trying to make a deep bow to true selfhood as a critical quality or engine of everything human beings do, then you're automatically setting yourself against everything that is dehumanizing in the culture," he said.

In the years he's spent carrying his question—"What on earth are you doing with your life?"—he's also come to understand something about Rilke's directive, especially when it comes to other questions that remain unanswered in Palmer's life. "Maybe, with the big questions, you won't ever 'live your way into an answer,' and you have to be comfortable with that," he said, paraphrasing the poet. "Rilke never issued a guarantee," he chuckled. "He said 'Perhaps you will then gradually, without noticing it, live along some distant day into the answer.' That's good enough for me."

As my visit with Palmer went on, I learned that laughter was the soundtrack of the house, something that Sharon, his second wife of thirty-two years, confirmed when she came downstairs to join our conversation.[19] The Minnesotan confessed that Palmer unleashed the comedian inside of her when the two met at a retreat center in the late 1980s.

"My family didn't laugh a lot," Sharon, who is a talented painter and quilter and Parker's first-line editor, recalled. "It wasn't that they were the Grimskies* or anything like that, but this was Minnesota, you know."

"You were just trying to survive in the cold," I joked.

"God's frozen people!" interjected Palmer.

"I learned a lot of humor from him," Sharon said, looking over at Palmer. "It was always there, I guess, but I needed someone to bring it out in me."

"We've laughed a lot from the very beginning," Palmer said, describing it as "a big tool for negotiating life" and its uncertainty. "There are things to which the healthiest response is laughter. It becomes a mode of knowing. There's an ultimately comic dimension of the human dilemma."

Parker and Sharon have used laughter to help them navigate the ambiguity of stitching together a blended family, to sustain two lives

---

* Sharon explained that this was a term that she made up to make light of her family's lack of laughter.

dedicated to creative and largely independent work, and most recently to "explore the territory called old age."

An example had come a few days earlier, when the two were out on their daily afternoon walk. "We both have bad backs, so walking is a little harder these days," Palmer said. "We can't walk as far, but we both work on that. She exercises, and I don't know what I do. I pray, I guess?" he laughed.

"We were out there with our aching backs, and we started laughing about a time years ago when we were in the woods somewhere and suddenly, without letting me know she was going to do this, Sharon ran at me from behind and jumped on my back." They were both smiling and laughing at the memory. "I'm six-four and was reasonably strong back then. But I looked at her when we were on this recent walk and I said, [referring to his aching back], 'You know, this is all your fault.'"

"At the time, I think you said to me, 'Do you love me or are you trying to kill me?'" Sharon recalled.

"I wondered—is this a horror film?" Palmer said.

"It was just a passionate instinct!" Sharon laughed.

Today, their laughter is not just a part of happy moments. It's also woven into "conversations about aging, uncertainty, and death," Palmer said. "We talk about that a lot."

"I can't imagine not talking about it," Sharon said.

"But in this country, there's a lot of not talking about it," he pointed out. "It makes my life so much easier to live with someone who isn't afraid of the questions, but is drawn to them."

Early on, they were drawn to each other not only because of their shared values, but also because of their differences—differences that sounded familiar to me. "One of the things I've wrestled with is that I tend to move quickly, and you tend to move slowly," he said, looking at Sharon. "For as long as I can remember, I've been out there flying around in imagination, experimenting, pushing the boundaries. I've had to learn more of what it's like to move slowly. It comes more naturally to me at age eighty-four than it did at fifty-four or thirty-four.

But a lot of what Sharon has brought to my life has been a grounding instinct, a grounding capacity."

And then, Palmer said something that made me see my own relationship—with myself, and with A.—in a new way.

"In a good relationship, you find in another person something that's missing in yourself," he said. "But ultimately you come to realize that you can't depend on the other person for that thing. You have to find it within yourself."

Relationships, he explained, can hold a mirror to the place where we most need to grow. But a relationship cannot replace that growth. Healthy relationships are not ones where each person completes the other, to paraphrase Tom Cruise's famous line in the movie *Jerry Maguire*.[20] Rather, they serve as a guide to your own self-discovery, helping you to grow more fully into who you are. A person cannot be our answer; but the right relationship can help us uncover what questions we should be asking ourselves.

"A lot of people stop at the point at which they've found something in another person that they lack, and they feel complete," Palmer said. "But they aren't. And if something happens in the relationship that stresses or strains it, everything feels desperate because they feel like they're losing this critical part of themselves. But the root of the problem is that they misunderstood the whole deal."

When we act as if our partner's purpose in life is to fulfill us, this not only stunts our own growth, but theirs, too. If they fulfill us just as they are, this means we'll feel threatened by any signs that they want to change, and we may try to shut that desire to change down.

In his relationship with Sharon, Palmer said, he had to learn how "to ground myself, which is ultimately the only grounding that works. I had to go through a period of not depending on her to bring that to me. Instead, I had to learn from the way she does that, and incorporate some of that in my own life, in my own practice, my own daily experience." He had to learn how to grow roots to complement his wings.

This, too, was part of my struggle with A. As I discussed in previous chapters, we often fought about living our lives at different speeds, with him complaining that I moved too quickly, and sometimes carelessly, through the world and me feeling restrained by his admonishments to slow down. But what I'd been missing was how much I often relied on him to keep me in check, how subconsciously I didn't think I really *needed* to change, because *he* would tell me if it was necessary. It was easy now to see how this was not only exhausting for him but also lazy of me. The point was not to define myself in opposition to him. It was to stop thinking about our relationship as consisting of two opposing, or complementary, parts. The point, I realized, was to start thinking about each other as whole people, separate from each other and guiding each other toward that wholeness.

Though I *knew* I needed to work on moving more slowly, more patiently, Palmer's articulation allowed me to see how this was essential not just for my personal growth but for the growth of my relationship, and even for A. I could not keep relying on him or anyone else to ground me. I had to learn how to ground myself.

To learn how to endure "within one's own I," in Rilke's words.[21]

I Take Thee, Questions

This was a lesson that photographer Marianna Jamadi had to learn, too. She discovered that committing to her questions allowed her to develop a sense of trust in herself that eased her transitions into and out of the most important relationships in her life. I found Jamadi and her art when I was searching online for people who appeared to have a deeper-than-average connection to Rilke—especially to the idea of loving the questions. At the time, Jamadi had developed a website for a side project she called *Making the Questions*. Though it was linked to Rilke's influence on her life and a dream she'd had about her dead grandmother whispering that phrase—"You have to make the questions"—to her, she wasn't quite sure what the project was yet.* I got in touch with her to find out.²² I ended up learning much more, starting with how she first began to design her life around the idea of living and loving her questions.

It all began at a time when she should have been deliriously happy. She was sitting on a Portuguese train, gazing at the countryside and planning her wedding. But all she felt was the persistent throb of doubts. On paper, her love story was the stuff of romantic comedies. Jamadi had met her fiancé four years earlier on a photo shoot for a fancy mall in Palm Springs. He was the photo assistant, and she was the assistant to a fashion stylist.

They had a lot in common. They'd both grown up with immigrant parents. They were both ardently creative. Jamadi was eager to hone her photography skills. She'd taken one class in college, but her partner was a professional, and she admired his craft. "A lot of times in the relationship, I had an idea and he could execute it," she said. "He ignited some kind of freedom of expression and my confidence in it. He allowed me to explore, but I also felt like I could depend on him."

He taught her technical skills, and together they built a wedding photography business, spending their weekends documenting the moments of love, joy, and comedy that live inside those big days. For a while, the work was thrilling, as they developed their unique style and the business grew.

But outside of their creative work together, Jamadi found it hard to connect with him. She recalled a dinner when she decided she would try not to talk about work, "and there was nothing to talk about."

And then there was this trip to Portugal. They were taking the train to meet his family, and from there they were planning to take a road trip to visit wedding venues.

But here she was, questioning whether her fiancé was the right one for her and whether the marriage "was going to leave me with more periods than question marks," she recalled. "It felt like I had so

---

* Ultimately, she ended up shutting down the website and pursuing other project ideas.

many more questions to live before I could arrive at answers or life commitments."

And there was an even bigger question on her mind: How could she trust her intuition if it was her intuition that had gotten her to this place? "I wasn't just questioning the engagement," she told me. "I was questioning my life's trajectory!"

Before her trip, she picked up *Letters to a Young Poet* to read, though she can't remember why. She *does* remember that Rilke's words—"live the questions"—immediately calmed her on that train. "I had so many questions, and I had been trying to answer them immediately," she said. "With that particular book, and that particular phrase, I had an 'aha' moment where I realized, 'Okay, I can't answer this quickly.'"

Jamadi started to live into her answers on the road trip to visit wedding venues. To her, living into answers means attempting to "inhabit different possibilities" for her life, trying to put herself in situations that allow her to sense what a particular decision might feel like or be like. The trip began inauspiciously—with a stalled emotional connection. She'd imagined the car rides would be filled with deep conversations about their life ahead, but their discussions were limited to trip logistics. "I would bring other topics up, but either he wouldn't know what I was talking about, or he didn't really interact," she recalled. "I felt a dissonance in terms of what we were interested in and knowledgeable about, which made conversation difficult. I was weighing the questions 'Can I live without that?' 'What's more important to me?' 'How much do I want to compromise?'" This disconnect wasn't new, but it had been easier to ignore when they were working together or talking about work—in other words, most of the time. Now, there was nothing to distract them.

At one point during their road trip, as they sped along the highway, it started to rain. Jamadi wanted to talk to her fiancé about how she was feeling, the gap she sensed between them. "I started to bring it up, and he was like, 'I don't feel that way,'" dismissing the problem. As they were talking, the windshield wipers began slowing down. The headlights began flickering. And suddenly, it wasn't just their connection that had stalled. The car died. They sat on the side of the highway, waiting for the tow truck for hours. By the time it finally arrived, it was 11:00 p.m. Jamadi remembered sitting in the truck, sandwiched between the operator and her partner, traveling to his parents' house, and "being like, 'I hear you, universe!'" She laughed.

When she left Portugal, she "kind of knew" that the relationship wasn't going to work, but the path forward wasn't clear. "We were so entangled and had a wedding shoot that was upcoming," she said. She wondered whether and how she could pursue photography without him and felt herself oscillate between fear and excitement, certainty and doubt. "When you're questioning everything at once it's terrifying and also so ripe with possibility," she recalled.

As she carried around her questions, they began couples therapy. For Jamadi, this made the decision to end it even clearer. "It was hard because it wasn't like anything terrible happened," she recalled. Quite the contrary—with him, she felt "more artistically free than any [other] relationship as he did unlock a part of me in that way. But that wasn't enough to stay in it. He just wasn't my lifetime partner."

She did, however, discover a different kind of partner in the form of her questions. With them, she'd found a new kind of relationship and a new kind of practice. After the breakup, another question began to tap at her door.

———

When Jamadi was eight years old, she recited a poem she'd written at the school talent show. The last lines were "I don't want to be a doctor, / I don't want to be a lawyer, / I just want to be an artist." She was wearing a red beret and a painter's smock made by her mother.

Though her parents humored her eight-year-old ambition, she knew they'd never support that dream. Jamadi's father was from Indonesia, and her mother was from Finland. "Both of my parents were immigrants, so there was this idea that creativity is nice, and it's a hobby, but it's not a career," she said. "I felt that first-generation pressure."

In college, she majored in business but took a photography class on the side. "I loved it," she recalled. "I started seeing the world through photos." The wedding photography side business Jamadi had created with her ex-fiancé had allowed her to survive a "soul-crushing and awful" corporate job at Victoria's Secret in New York City. But without their business, she found herself back on the sets of photo shoots, steaming clothes, and more interested in who was photographing the models than anything else.

She began to leaf through her old dreams. What would it mean for photography to be her full-time job? Could she do it alone? "It was hard to even ask myself, 'What do you really want to do?'" she said. "But Rilke's words gave me permission to ask it."

Asking the question allowed her to see the path to answer it: What she really wanted to do was to be a travel photographer. But how could she know whether this was the right answer without attempting it? In 2013, Jamadi embarked on an experiment: She sold her belongings, quit her job, and began a year of traveling designed to help her learn "what it meant to be a photographer out in the world, and to be a travel photographer." But she was also exploring an even bigger question: "Who am I—as an artist, as a person—without my ex-fiancé?" She had to learn how to trust herself and to begin paying attention to what felt right to *her*—not to anyone else.

Leaving her smartphone behind, she made her way from Los Angeles to Barcelona, Spain, and then to sixteen other countries. She snapped photos of whatever inspired her, not thinking too much about it. She allowed herself to feel "major and minor at the same time," a tiny speck next to the ocean and yet a part of all the vastness. Traveling built her confidence, allowing her to see that she could navigate the world on her own. After the trip, she had not only a better sense of her artistic voice, but also the answers to her questions.

"The answers were, 'Yes, you can do this, and yes, this is you,'" she said. "So when I was done with that trip, I didn't even give myself the option to go back to any other job." Since 2014, she has been a full-time freelance photographer.

As her creative work began to blossom and she landed her first big gig—photographing a luxury hotel in Mexico—another period of uncertainty hit, threatening to unmoor her: the death of her father and mother. They died in back-to-back years, between 2017 and 2018.

"I didn't know how to handle it," she told me. "I've always created out of those spaces, out of the questions: What is this? How do I deal with it? How can I get through it? Are we still connected? Death, and mortality, [raise] unanswerable questions. It can be a reckoning experience. Things that felt solid become gelatinous, and you're kind of grasping."

It felt, Jamadi said, like a "break in the timeline—like there was a me before, and a me after." In some ways, she felt she needed to rewire her brain to break old habits—including the way she used to always text her mother after she landed in a new city. For six months after her mother died, Jamadi would reach for her phone, only to remember that her mother wouldn't be there to receive her message.

Her identity was changing, too. Her parents' death made her feel more distant from her cultural heritage. "I was so sad to think that my [future] kids will never hear the languages they spoke in the way I did," she said. "They'll never meet them. I felt robbed of all these things in my future."

Death felt like the ultimate permanence, the ultimate distance. And so, at first, she sought connections with others. Jamadi surrounded herself with older creative women, befriending a group of women in Los Angeles who had created a circle for artists. These women helped bring her back to her artistic intuition and instincts at a time when she felt shaky. As she created, she continued to carry her unripened fruit questions: *Can I still be connected to my parents? How do we remain connected? Is it really over?*

As a travel photographer, she'd witnessed the way other cultures created meaning from their grief. Each day, around a hundred bodies are cremated on the banks of the Ganges River in Varanasi, India; Hindus believe that this rite marks the end of their reincarnation cycle and allows them to reach nirvana.[23] "Death is a very present thing in that city," Jamadi said. "There are constant cremations and chanting.

In the moment, I felt like, this is really uncomfortable." But when her parents died, Jamadi reflected, "I found myself trying to reach for experiences like that, that felt like there was more of a continuation [after death], more meaning."

She wondered whether her relationship with her parents could punctuated with a comma or an ellipsis, rather than a period. Both had been cremated, and she wasn't sure what to do with their ashes. The urns she saw were all ugly; as an artist, she would never want to display them in her home. When she talked to other people who also had ashes of loved ones, she mostly heard that the ashes were in a box in a closet. "The Western world doesn't like to talk about death as much as other cultures," she pointed out. "It feels taboo." But she also noticed that most people weren't asking questions about whether they could rethink their relationship to their loss. For them, she said, there was a period at the end of these relationships. But "for me, I wondered if there might be more to the story."

Six years after her parents passed, she started to extend that story by creating urns with a secondary purpose: to help people create rituals to connect with deceased loved ones.[24] Her urns double as candleholders and vases, allowing owners to weave the memory of their loved ones into daily life. This way, Jamadi explained, owners can interact with their deceased loved ones through the objects—through the simple ritual of lighting a candle or putting the person's favorite flowers in a vase on their birthday. It is a way to "tend to the relationship" we have with our grief and with the ones we've lost, she said.

And this is where Jamadi discovered something surprising. In the years since her parents' death, her relationship to them has changed. Take, for instance, her relationship with her father, whom she described as "a very mysterious man. He was not a very emotionally expressive person. And I always felt like he didn't understand or support me as a creative."

After he died, she began to sift through his things, what he chose to collect. "It's a lot of stuff that surprised me," she said. These things include an entire folder of articles and papers detailing Jamadi's artistic accomplishments, as well as the photos and videos he took on trips. Her father, as it turns out, was an amateur photographer.

Watching the VHS tapes "is the closest thing to seeing the world through his eyes," she said. And what did he videotape? "It's always family and his kids. That was a way for him to connect when he didn't know how to emotionally connect. Maybe he felt it was easier with a camera in between us."

The Shape of Questions

In the aftermath of her grief, Jamadi continued to hone her question-asking practice. It had become a way for her to move through the world, a tool for her to come back to who she is and to discern who she is becoming. "It has become part of the rhythm of how I live," she told me.

She relied on it again when, in 2023, newly married, Jamadi began what for many women is another highly uncertain process: in vitro fertilization.[25] Women who undergo the arduous, emotionally and physically taxing procedure have often struggled for months or years to become pregnant through sex alone.[26]

During Jamadi's long months of hormone injections, egg extraction, fertilization, and waiting, she explored her questions through journaling and art. "Every day during the IVF process, I drew a line of rectangles on paper," she told me. "It was my way of meditating on my uncertainties and simultaneously creating some kind of structure, pattern, and rhythm of that time."

Jamadi isn't sure why she chose the rectangles, but thinks they may have come from "an unconscious feeling that they feel nice and contained when I was feeling out of my mind," she wrote in an email. "I also liked how with the drawings, I could tell where I was at that day. Sometimes the rectangles were nice and straight, other times shaky and uneven. It was a good practice of actually SEEING how I was feeling that day."[27]

After two rounds and two transfer attempts, Jamadi got lucky:

She became pregnant with twins. One question—Would she ever get pregnant?—was answered, just as many more bloomed about who she would become as a mother.

Letting the Fear Catch Up

From the outside, Tom Rockwell's life looked to be exactly where he wanted it to be. He was married with young kids. He'd built a flourishing career, first founding a company that designed museum exhibits and fabrication, and later becoming the creative director at the Exploratorium, San Francisco's museum of science, art, and human perception.[28] At the museum, there was always someone to talk with, a new project to dream up, more work to create. Rockwell was so busy that he was turning down meetings with Nobel laureates.

Meanwhile, he harbored ambitions to someday become a museum CEO.

He regularly pulled all-nighters, intoxicated by the adrenaline of creative productivity, of ephemeral accomplishment. But afterward, he'd feel a sense of dullness as he came down from the high. The cycle repeated.

Then came Christmas of 2005, one year after he started his job at the Exploratorium. Rockwell had taken a rare week off, and on the last day of his vacation, he picked up a book at an Oakland bookstore about the secrets of highly effective CEOs. He waited to read it until the night before he was supposed to go back to work.

According to the book, the most in-demand CEOs received—and were expected to stay on top of—close to five hundred emails a day. As Rockwell turned the pages, he grew more and more anxious about this idea. As it was, he could hardly manage one hundred emails a day. How would he filter through five hundred? And was that what he really wanted to be doing? "That was a pivotal moment," he said. "I began to realize the path I was on was only going to lead to more and more emails."[29] He also began to see something else—that his relationship to work was a form of addiction.

That addiction had consequences. Beneath the surface of his success, his marriage was dissolving. His kids were in distress and struggling. His relationship with work was a common topic of discussion with his then-wife, "especially when her grandmother was dying and I wasn't able to be sufficiently present for her," he told me.

But his was not the average tale of workaholism. The conditions for it were established decades before he was born.

The Other Rockwell

Today, it's easy to find images of "ordinary" life on social media—of someone's breakfast, their morning workout, a dinner party, a family raking leaves. But before these portrayals became ubiquitous, Norman Rockwell's paintings played a similar role, proffering portraits of regular people's lives.[30] Norman Rockwell was Tom's grandfather, and in mid-twentieth-century America, Rockwell's early illustrations offered wildly popular sketches of life and culture that were often sweet, simple, and racially homogeneous: a family gathered around the table for Thanksgiving dinner, a kid's first haircut, children spellbound by the moon.[31] Rockwell's art captured what many critics argued was a too-sentimental, saccharine image of American life.[32] Meanwhile, his fans adored the artist's ability to honor the details of their lives during the tumultuous period that included two world wars, the Great Depression, and wars in Korea and Vietnam. President Dwight Eisenhower commissioned Rockwell for his portrait, as did Presidents John F. Kennedy, Lyndon B. Johnson, and Richard Nixon.[33]

Like today's social media, Rockwell's early art served the public slices of a curated life, often in the form of the covers he painted for *The Saturday Evening Post* between 1916 and 1963. In 1916, in the absence of radio, television, and the internet, "the *Post* was the national frame of reference," explains Deborah Solomon in her biography of Rockwell, *American Mirror*. "As the largest-circulation magazine in the country, it helped spawn a new kind of culture: mass culture."[34] Its millions of readers could compare their own messy lives to these intricately detailed portraits, taking comfort in their similarity or fretting about the distinctions.

Norman Rockwell's personal life didn't look like his paintings. His first wife, Irene, divorced him, and his second wife, Mary, suffered from depression and alcoholism. Rockwell, too, suffered from depression.[35]

Eventually, Mary became a patient at a psychiatric center in Stockbridge, Massachusetts, where Rockwell moved his family and began receiving treatment himself from the German-born psychoanalyst Erik Erikson, who is known for bringing us the term "identity crisis" and elucidating the stages of human development.[36] Erikson joked to Rockwell that his family had "logged more hours of psychiatric care than any other family in America."[37]

Behind Rockwell's affable public mask "lay anxiety and fear of his anxiety," writes Solomon. "On most days, he felt lonesome and loveless. His relationships with his parents, wives, and three sons were uneasy, sometimes to the point of estrangement. He eschewed organized activity."[38]

According to journalist Tom Carson, Rockwell was "at ease only when at his easel," taking "little interest in hobbies—or even in his family."[39] When CBS's Edward R. Murrow asked him to describe what he did in the evenings, after the work of the day was through, Rockwell gave him an example of spending his time tearing up diaper cloths that he used as paint rags.[40] Rockwell was obsessively, neurotically clean—scrubbing each of his paintbrushes with Ivory soap. He once quipped that someday he might spend his time *only* cleaning up his studio, rather than painting.[41] He celebrated Christmas by working only half a day.[42]

Work was his purpose and his distraction. While his second wife, Mary, continued to abuse alcohol and pills, Rockwell "took refuge in his studio, allowing the prod of deadlines to drive out rogue thoughts," Solomon writes.[43]

The message of the Rockwell home was clear: Life was work.

And it wasn't just work that was important. It was talent. You had it or you didn't.

All three of Rockwell's sons had "it," though Norman disapproved of his son Peter's chosen vocation of sculpture. "That's nonsense," Norman told Peter. "Jarvis is a painter, Tom is a poet, and the only thing I can think of that is commercially worse than painting and poetry is sculpture."[44] But Peter couldn't stay away from his clay, bronze, and stone. After college and a stint at the Pennsylvania Academy of Fine Arts, he and his wife moved to Rome to study the techniques of the Italian Renaissance sculptors, and they raised their four children there.[45]

One of them was Peter's son, Tom Rockwell.

As a kid, Tom was aware that "not all was well in the kingdom of fame," but he was too young to understand the complex dynamics at play in his family—the mental illness, the way power distorted their lives, the societal pressure and expectations.[46]

What he did know was that being part of a famous family was pretty cool. "You feel it, you notice it, as a kid," he told me. Any American tourist who heard his last name would ask, "Any relation to Norman Rockwell?" And Tom would smile, confirming that there *was* a relation and basking in their subsequent delight. Sometimes, he'd try to engineer conversations so he could bring up his famous grandfather.

He saw his father derive influence from the fame of *his* father, becoming a successful sculptor whose work is displayed in St. Paul's Church in Rome and the National Cathedral in Washington, DC, among many other places.[47] And Tom felt the power, too.

He wanted more of it. And the path to power seemed clear: natural artistic talent.

"The way [my dad and siblings] would talk about the news, the way we would have arguments, there was a lot of emphasis placed on being smart or talented," he recalled. It wasn't just the desire for fame. The goal was to earn status or power from cultural accomplishments, as his grandfather had. "The idea of becoming a successful businessman—I was totally not interested in that," he said. "It was all about becoming famous for something I've done that's cultural."

He began to search for what that something could be. At six years old, he started acting in ads for G.I. Joe and Lite-Brite, and then in high school he was an extra in the movie *Luna*, directed by Bernardo Bertolucci.[48] But the theater community, he decided later, felt too shallow.

As a high school student, he also took art classes. But these classes didn't feel right either. He concluded he wasn't talented enough and didn't have his grandfather's eye for representational work.

Then, in college, he found music. *This*, he thought, could be it. So he learned to play guitar and flute. Still, he doubted his abilities. He thought perhaps the problem was *where* he was living, so he began traveling back and forth between the US and Italy.

"I was stuck in perfectionism," he said, reflecting on his twenties. "I spent a lot of time being enamored with the dream of being a prodigy . . . if I just found the right thing." Rockwell, like Parker Palmer, was trying to live his life according to a distorted idea of who he *should* be, chasing what he thought was the answer to a successful life rather than exploring questions about what that life could look like for him. "The shadow of [my grandfather and father's hunger for] fame hung over the possibility of what I was looking for in life," Rockwell told me. "Fame is complicated because you have this sense that by making art you can become important, and all of these people will know your name."

When Tom's father, Peter, was a teenager, he spent a year in Hollywood while his dad was working on paintings there, later regaling Tom with stories of 1940s movie stars. One of the stories Peter told again and again was about a special guest who showed up one day at their kitchen door in Arlington, Vermont. The guest asked Peter if he could see his father, Norman. Peter asked, "Who should I say is calling?" The guest answered, "Tell him it's Walt Disney." It was a moment, Tom said, in which his father realized how famous *his* father was.

The shadow of fame haunted Peter, too. "My dad [Peter] was focused his whole life about whether he was a good, important, and worthy artist almost right to his dying day," Rockwell said. "I would often find myself reassuring him that he had done well and should be proud, but the issue would pop up again and again."

From Perfectionism to Workaholism

By his early thirties, Tom Rockwell had come to a painful recognition: Perhaps he wasn't an art prodigy. Around that time, he'd taken a creative risk, developing and starring in a one-man theater show that grappled with questions about science and religion. This topical intersection fascinated him, and he felt others were not yet trying to explore that intersection through art. So he poured his soul into developing the show, which was in three parts. The first told the story about a mask maker for the gods who wakes up with a vision on the eve of the scientific revolution: In the future, there will no longer be a need for gods with faces. The second part was about a mystical force that came to steal mathematics from humanity. The final part was about monastics in the future who had developed a ritual that allowed them to consider the size of the universe through movement and sculpture.

When Rockwell finally performed the show in Ithaca, New York, and in New York City, he received polite applause. For him, this tepid reception was devastating. Instead of spending time in the uncertainty of what it all meant, he took it as a sign that he should enter a more predictable career. Perhaps creating art wouldn't be his path to power and importance. But that didn't mean he couldn't work his way to prominence. When he later became a museum exhibit director, his perfectionism continued to morph into workaholism.

Slowly, steadily, the workaholism drove wedges between him and the people he cared about. He was in a race that he could never win, continuing to hold on to an idea, an expectation, that he would become famous like his father and grandfather. But he wasn't his father or his grandfather. The reality was that he was "clinging to notions of self and trying to get them to last longer than they might last," he explained. The person he thought he should be was not the person he really was.

This reality was terrifying—so terrifying that for most of his life, he'd been trying to avoid this truth and the questions it raised about who he was, how much work really mattered, whether he was worthy, whether he was okay. "Fame and fame-seeking became substitutes for other types of spiritual work, and that was the case for me and my relationship to work," he reflected. "I was seeking something. And seeking is motivated by questions, whether they are explicit or not."[49]

For him, the experience of addiction represented an attempt to find a shortcut to an unanswerable question. At the time, he described the questions running his life as "primitive. How do I do more? How do I be special enough so that I don't get in trouble? How do I not be terrified?"

After the Christmas when he saw his relationship with work for the addiction that it was, he heard about a group to support workaholics that met within walking distance of his house in Oakland. As he met with the other participants and set boundaries around his work, he realized he would also have to dismantle and change the questions that had been fueling his workaholism.

As soon as he stopped trying to manage his terror through overworking, "the next question was: What is this fear? How do I address this fear differently? Suddenly I had to look more at that fear, to sit with it. I had to stop medicating my own anxiety and had to stop feeding those questions with overwork."

What *was* his fear? He was afraid that he would be socially rejected at work, or in life, if he failed to live up to a certain standard. But as he stuck to his new boundaries—working only forty-five hours a week and no more than eleven hours in a single day—something surprising happened. And that was . . . nothing. He didn't get fired. His colleagues didn't lose respect for him. No one seemed to care.

By living out the question "What if I don't work all of the time?," he was giving himself evidence to refute what he'd always imagined the answer would be and was starving his fear with new lived experiences.

"For a lot of my life, I could outrun the fear," he said. But as he outran it, he also sped away from opportunities. He wished, for instance, that he had become a scientist, a path he'd stayed away from because "I was so scared of not having an answer or not having the talent that I never allowed myself to hang out in the questions."

And he found another strategy to deal with his fear: asking different questions. So far, he realized, the questions motivating his life weren't helping him get the sense of fulfillment and purpose he wanted. One of the questions that replaced them was "How can I help others?"

As he continued to reflect, he uncovered the question at the root of all his other questions: "Do I matter?" Rockwell realized he had to be "as comfortable with death and insignificance just as much as I am with significance. [Chasing] fame was an attempt to try to cheat death."

The questions of death, significance, mattering—all are, at some level, unanswerable, he concedes. Some of these are the heartwood questions that stay with you throughout your life, evolving alongside you. Others are questions that, like dead leaves, you may be better off releasing. A big part of his recovery was reckoning with the delusions of grandeur, of significance. Even questions like "What am I *meant* to be doing, what is the true me, what is my calling?" are part of his addiction, as they create a picture of a spandex-clad, never-changing hero, rather than a messy, dynamic, contradictory person.

Today, he still nurtures big ambitions, but he tries to "hold those lightly enough so that I'm not living some fiction. I'm not arguing against reality with my life."

He has now been in recovery for sixteen years and recently saw just how much his relationship to uncertainty and questions has changed. Even after the devastation of his one-man science and religion show, he had continued to cultivate an interest in those fields. Decades later, he began finding ways to come back to the work. He started a project developing museum exhibitions exploring the intersection of science and religion, and was asked to present his ideas at a conference. "I knocked myself out to do a presentation for [the conference] and put a lot of myself into it, similar to the one-person show," he said. And like the one-person show, people responded with muted positivity. "It wasn't a failure, but it was clear that there was going to be a lot more work to do," he said.

Afterward, Rockwell felt some familiar stirrings to "crawl back into the cave," to foreclose an answer or conclusion about what it all meant. "Part of me wanted to abandon the whole project," he said. This time, he tried something different. He took some time off to rest in the questions. The public reaction, he realized, "may have nothing to do with whether I delivered a great presentation or not; it may have nothing to do with whether I have a lot of work left on the exhibition or not. It doesn't have to be a big story about 'I'm a god' or 'I'm a worm.' I kind of just knew that I needed to rest and not come to any conclusions. It's okay to be a godly worm."

Rockwell also remarried after a fifteen-year relationship with his now wife. They spent their honeymoon at a weeklong silent meditation retreat, which for Rockwell was a surprisingly intimate experience. The silence, Rockwell said, gave him a chance to get curious about his wife, about marriage, about love and how it can "become something different all the time." When you're with someone for many years, he said, "you do have to reinvent your relationship. Getting married was an example of that, and the silence created more room for this idea that it can be a lot of things, and we get to decide."

―――

What does a life of loving our questions look like? Rockwell, Jamadi and Palmer showed me that it is a life spent pursuing a closer relationship with ourselves, asking ourselves questions, listening patiently to the answers, and considering what those answers might mean for our life course. From them, I learned that questions could serve as a kind of internal GPS—a way back to who we are and what we need during moments of uncertainty.

At first, this point may seem counterintuitive: Questions, after all, *appear* to be leading us someplace unknown or wholly new. But what I found is that the spaces they lead us to are both novel and familiar at the same time. We haven't been there before, and yet they feel like home—kind of like the sensation you sometimes get when visiting a new city: *I could live here.* That's because questions lead us to places that exist already inside of ourselves, discovered long ago or waiting to be found.

Our relationship to ourselves, then, is the first and most important one to cultivate to pursue this alternative relationship to uncertainty. But I also heard something else in the stories of Palmer, Rockwell, and Jamadi, as well as the stories of others interviewed in previous chapters: the role that communities and relationships played in helping them sustain lives lived in questions.

How much to rely on others during times of uncertainty, however, was a delicate balance for each of them, just as it has been for me. During my most desperate moments, I've been tempted to start seeing other people—therapists, friends, and yes, astrologers—as not just systems of support but as purveyors of answers. Rilke issued a warning about this strategy, proclaiming that it could "lead to nothing but great difficulty."[50]

I knew I needed to better understand how other people were using community in their quest to live better with uncertainty. Soon, I discovered that this would mean forever changing the way I defined community and relationships.

5

How Do We Fall in Love with Questions?

THE BIG IDEA

The most important part of being able to fall in love with questions is to surround ourselves with nourishing relationships and communities. For many of us, this part of the journey entails acknowledging a paradox: that to find the right community, sometimes we need to let go of what we think the right kind of community is. Communities need not always be composed of people. We can find community in books, nature, and even our own questions.

It is morning, and Rilke knows he needs to work. He has donned a tailored suit and tie. He sits at a desk decorated with a silk scarf and flowers. He must write. Time to start. But as he picks up his pen and stares at the blank page, he feels a tug—not toward his poetry, but something else: other people.

"In addition to my voice which points beyond me, there is still the sound of that small longing which originates in my solitude and which I have not entirely mastered," he writes in a January 1920 letter. It is "a whistling-woeful tone that blows through a crack in this leaky solitude—, it calls out, alas, and summons others to me!"[1]

Solitude is one of Rilke's favorite subjects and one of his favorite experiences. But even he, whom W. H. Auden called the "Santa Claus of loneliness,"[2] does not fester in it, and sees the way community and solitude nourish each other.[3] "The more solitary a person is," he writes in his "Notes on the Melody of Things," "the more solemn, moving and *powerful their community."*[4] *The more one gets to know herself, the more she can connect to other people.*

Rilke cannot always venture out to see others and often does not want to. And so, instead, he writes letters—thousands of letters throughout his life. Kappus is just one of hundreds who correspond with Rilke and puncture his solitude.[5] *These people are a vital part of his community.*

Like Ralph Waldo Emerson and Henry David Thoreau, Rilke sees fleeing from people, industry, cities, as a way to uncover truths inside himself that could connect him to others, writes Rilke translator Damion Searls. "They found spaces to retreat to . . . and they did so, yes, to avoid other people, but also to reach them."[6]

Whenever I tell someone that I'm an only child, their reaction is almost always the same.

"Oh wow," they say, always a little surprised, as if alighting on a rare species of tree frog. "Were you lonely growing up?"

Was I lonely growing up? Since I was a kid, loneliness has fluttered in and out of my life, a presence that flits, hovers, dives, and soars. It is sometimes a hawk, with a wingspan that temporarily blocks the sun, circling ominously overhead. It is sometimes a bat, emerging at night in shadowy corners of places that once felt safe. But other times, it's a butterfly—an aching solitude that inspires lightness and creativity.

When I was six, I wanted a sibling so badly that I fabricated the existence of one. To anyone who would listen, I explained that I had a sister who went to work with my father every day. This was the reason no one would ever see her at school or other activities. Once, when my mother picked me up from school, the teacher asked her about whether we'd ever get to meet my sister. My mom stared at her, and then back at me, before exposing my lie.

Being alone forced me to rely on myself for entertainment. I spent afternoons writing short stories, reading, drawing, or painting. All this time alone also made me . . . kind of weird. As just one example, I spent days writing and illustrating a book that documented the secret lives of the squirrels who lived in my backyard. I'd named them all and developed their elaborate backstories. I also enjoyed feeding them peanuts—from a distance, of course, except for the time one ran up to me and grabbed a nut from my hand.

This weirdness, unsurprisingly, made it harder to find friends.

Eventually, I found one. For two glorious years, she and I did everything together: rode our bikes, had sleepovers in her backyard, managed successful lemonade stands, built pillow fort palaces. Her family was big and raucous, a stark contrast to my small, quiet one. Spending time at her house felt like stepping foot on another planet. It was closer to what I imagined a Normal Family was like, as if she existed in a Norman Rockwell painting and I in a Vincent van Gogh.

This one-best-friend strategy worked—until it didn't. When we were in fifth grade, she ditched me for another girl. It was a mid-1990s version of ghosting that was perhaps more painful since we were still in the same class, and yet she suddenly had no time to spend with me. At any age, this is devastating. But at ten, it felt like the end of the world—especially for an only child like me. I was already worried that I would never have the kind of relationships people seemed to have in books and movies.

As I got older, I started to believe there were two parts to my problem. First, if I wanted more friends, I had to become cooler, more like everyone else I saw around me. And second, I couldn't invest too much in any one relationship or group. I had to diversify my friendship portfolio.

Slowly, I shed what I perceived to be some of my uncool traits: I traded my glasses for contact lenses, grew my hair out so it was less umbrella-shaped, got my braces off, and started dressing in Abercrombie & Fitch attire. I stopped watching, and singing along with, Gene Kelly musicals and began memorizing Britney Spears lyrics. Even if I still couldn't flat-iron my hair perfectly, or at all, at least I was flat-ironing the spirals of my weirdness.

As a freshman in high school, surrounded by nearly four thousand new potential friends, I felt something start to click. Miraculously, people seemed to *like* me. Or at least they liked the version of me I'd become. I became obsessed with accumulating as many friends as possible and becoming a member of as many groups as I could. Maintaining lots of superficial friends was my way of hedging my bets, staving off my vulnerability to rejection and hurt. I was trying to guarantee that, if one group fell through, I'd still have another one. I carried this mindset and friendship strategy with me into college and then into young adulthood. Throughout my twenties, I was often frazzled, running from party to party, returning at the end of the night feeling exhausted and hollow. I'd never really been present anywhere, rather I was always looking at the clock to see how long I could stay before the next engagement. I was in communities without really being *in community*. Because I couldn't yet feel at home with myself, I couldn't feel at home, either, with other people. What was I searching for? Was it just friends or something more? Something different?

As I spoke to dozens of people who were trying to renegotiate their relationship to uncertainty, I heard again and again about the importance of community and relationships to the quest of learning to love the questions of our lives. And this made sense, considering what community evolved to do: enhance our sense of security during uncertainty. Hundreds of thousands of years ago, early humans realized that by pooling resources and mental capacity, they had a better chance of surviving in the wild.[7] Modern neuroscience shows us that a sense of being connected in a community not only reduces stress-inducing cortisol but also activates the parts of our brain that indicate we are getting a reward and we are safe.[8] The right kind of relationships and community have the power to help us feel secure even while we hold questions that make us feel the opposite. This puts us in an optimal state of mind to explore questions, rather than rushing toward answers.

Hearing repeatedly about the importance of community forced me to examine my relationships. Though I'd belonged intermittently to communities in the past, I did not belong to a community designed to help people love questions. I imagined a group of close friends and family in my living room, coming over for regular exchanges about the big questions we were carrying, the uncertainty we were facing at that moment. I wondered: What did communities explicitly tailored to help people live in uncertainty look and feel like? What could I learn from them, to take into my own life?

Luckily, I wondered aloud to someone who could help: Casper ter Kuile, author of *The Power of Ritual: Turning Everyday Activities into Soulful Practices*.[9] He asked if I'd like to join Nearness, a community space he cofounded where people can explore life's big questions.[10] At that time, it consisted of peer-facilitated small groups that met each week for ninety minutes over Zoom for eight weeks at a time. Inspired by Ella Fitzgerald's version of the song "The Nearness of You,"[11] the name was also meant to suggest proximity to other people—but more of a comfortable closeness, rather than a smothering, on-top-of-each-other vibe. To me, it sounded a little cult-y, but I tried to stay open.

"People have fears about communities being overwhelming, and spiritual communities being domineering," Alec Gewirtz, Nearness cofounder told me, in a later conversation.[12] They wanted to set different expectations and to create a more structured approach. They've found it's crucial to do everything you can to *remove* uncertainty in these gatherings, so that people feel more comfortable bringing their uncertainty, and questions, to the surface.

I was sold when I read one of the featured blog posts on the Nearness website. It was titled "Is This a Weird Cult Thing? Or a Front for a Church?" (The thoughtfully written response: No.) Within the hour, I was signed up for the summer session, which started in just a few days.

What if We Already Belonged?

Gewirtz's path to Nearness started when he was hanging out with friends in grade school, playing games in a friend's basement. He saw his friend's cousin, who was on the autism spectrum, standing at the top of the stairs. The cousin, it seemed to Gewirtz, wanted to come downstairs and join the group. Instead, he stayed at the top of the steps, hiding behind the door.

"I remember the image of the person staring down at us, listening to what we were saying," Gewirtz said.[13] "It moved me, and I found myself wanting to learn more about the experiences of people with disabilities."

He began to work as a peer mentor in a social skills group for people on the autism spectrum when he was in high school, modeling what it was like to shake someone's hand and maintain eye contact, or to talk to someone you're interested in romantically.

"I would pretend to be the one who was good at all of these things, when of course I would also feel awkward as a teenager if I ever tried to flirt with somebody," Gewirtz recalled, smiling.

When he went to college, Gewirtz's interest in disability deepened, and he began to notice changes in himself that challenged his sense of who he really was. He began experiencing anxiety so debilitating it would lead to panic attacks. These were moments, he said, of "having my own weakness put in front of me," of losing his sense of autonomy, and understanding that "our bodies are unruly things that can go wrong in lots of ways."

Gewirtz began to ask questions like "How do I come to terms with weakness in myself and others?" This question eventually made him pursue a role as a caregiver inside the L'Arche community, where he saw "community at its best." There are more than 150 L'Arche communities around the world where people with and without intellectual disabilities live and work together.[14] Starting Nearness was partly motivated by wanting to bring those elements of community to other people.

Through L'Arche, Gewirtz learned about the "central importance of vulnerability" in community. Though we may feel more comfortable talking about what's going well in our lives, our relationships with other people get stronger, faster, when we bond over shared struggles.

He also learned what it meant to be present with someone else—not to try to fix them or give them advice, but to simply be there with and for them. He recalls lying on the floor next to a man who had just had a seizure in his bed and didn't want to be left alone because he was afraid.

And, fundamentally, he learned about forgiveness. In community, people can disappoint you. You can disappoint them. It's critical to create a shared practice to surface those disappointments or issues, along with a way to welcome people back in.

In our first Nearness session, we were given instructions to draw a wavy line on a piece of paper. One end represented our birth, and the other end was where we were today. Next, we were to document moments of connection and belonging above the line and times of disconnection below. Then, we shared what we wrote down. After each person shared, the rest of us would tell them what in their story resonated most with us.

For four minutes, I rambled through my life story of belonging, ending by sharing the reason I'd wanted to join this group: I felt I was lacking a solid community of people in my life who were willing to regularly grapple with big, meaningful, and unanswerable questions—the heartwood or pawpaw varieties. And I found myself hoping, strangely, that other people would say something similar. Or that I would see more nods and looks of recognition.

But other people didn't say similar things. Their stories, hopes, and expectations were different from mine. Had they all figured out this whole community and connection thing? Was I the only one feeling this gap? I was reminded of a conversation I'd had with Rachel Boughton, a Zen Buddhism teacher, about community. There's a warm and gooey side of community, but the other side is "that community can make you feel like you don't belong," she said. "Even if you felt close to people, community can drop kick you—one person says the wrong thing, or they start joking about something that horrifies you."[15]

We have to learn how to be in community, and this isn't easy—especially for the people who might be interested in communities like the ones I was seeking. "Most of us who are interested in big questions, we're not extroverts," Boughton said. "We don't easily fall into shallow conversations and loud parties." But we still need to "learn to listen, and part of that is really listening internally, giving yourself space when you need to, and not necessarily believing that if you feel uncomfortable in the moment, that you don't belong, or you aren't part of it." Though we often come into communities with a specific craving and an idea of what would feed it, there's also an opportunity to take everything as it is. "Maybe it will meet my craving if I let myself have all of it—the guy in the corner who is grumpy, the person over there who is super insecure and talking themselves up," Boughton said, imagining a fictitious party. "Most of the time I don't have to reject it and leave."

There was another way to look at the distinctions between me and the other community members, too. If, in fact, they had unlocked a deeper sense of community in their life, this was a good thing. This could mean that I had more to learn from them.

During the second week, we sketched a "wheel of belonging." We assigned each section of the wheel a category—like family or neighborhood—and filled it in based on how great a sense of belonging we felt to it. As we read through what we put on our wheels, I was struck by how much I wanted approval from this group and to be the one who was sharing the deepest insights. I was focused on how people responded to my comments versus others, and I felt as though I was making some of the *least* interesting contributions. I tried to stay curious about this feeling, rather than to chide myself. What would allow me to open up and simply listen and learn, rather than looking at this as an opportunity to burnish my reputation? And all of this for people whom I would likely never see again after those eight weeks!

At another session, we shared what each person had brought to the group so far. People told me that they appreciated my vulnerability and earnestness. *Earnestness?* I balked. I didn't want to be earnest! I definitely wasn't earnest. They clearly didn't know the real me. (Okay, maybe I was a little bit earnest.)

The fifth week, something shifted. We'd all been encouraged to bring some object from the natural world and ponder what it would be like if we became the item we brought. What new perspectives would it bring to our lives? My object was the leaf of my favorite tree—the ginkgo. One of the oldest tree species left on the planet, it even predates some dinosaurs. It's known as a "living fossil."[16]

We laughed together, marveling at how strange it was to stare at a leaf, a blueberry, a jar of sand, a rock, for more than a few moments. People started recalling experiences that group members had shared in previous weeks, contributing to a growing feeling of warmth and familiarity. I had to admit: They were starting to feel more like friends than strangers. Something was working, penetrating our collective armor.

Week six was, unexpectedly, only me and one other group member. We delighted in going "off script," talking about how taxing we found another Zoom meeting after a day of videoconferencing and why we kept showing up. She'd been looking forward to that week's discussion—focused on belonging to spirituality—because this was a space where she, an expert on machine learning, considered herself a little lost. When it came to "woo-woo" practices,* she told me, she liked to journey "one woo in," using tarot cards, for instance, to "introduce noise" into her decisions. Science, she said, didn't explain everything she had experienced in her life, and she craved more. We talked about questions, and she wondered whether questions, like fruit, had a particular time that they became ripe, ready to be picked and answered (an idea that helped me develop the questions framework described at the beginning of the book). She shared a beautiful poem with me—David Whyte's "Sometimes," which describes the kinds of heartwood- or pawpaw-style questions that "can make or unmake a life," the kind that "have patiently waited for you" or "have no right to go away"—and I decided that this exchange had made the whole experience worth it.[17]

---

* The term "woo-woo" refers to beliefs, ideas, and practices—often involving the supernatural or paranormal, such as astrology or crystal healing—that may be considered unconventional or pseudoscientific.

We were building toward a grand finale. In the final weeks, we asked, "What would change if we believed we belonged to everything—our neighborhoods, our families, nature, ourselves, the larger whole?" This question reminded me of a remark of Rilke's from 1898: "All disagreement and misunderstanding originate in the fact that people search for commonality *within* themselves instead of searching for it in the things *behind* them, in the light, in the landscape, in beginning and in death," he wrote. "By doing so they lose themselves and gain nothing in turn."[18] We obsess over our connection to other people, neglecting the ways in which we're connected to so many other parts of the world. What would change if we *already* saw ourselves as interconnected beings in community?

The group members talked about how this might mean they could live without listening constantly to their inner critic, that it might mean they could live without crushing expectations of who they needed to be *in order to* belong. It could allow them to trash the fear of "getting it wrong" they constantly carried.

Our final exercise together was to celebrate what each person had brought to the group. This was unexpectedly moving, as each person presented gifts of genuine compliments, memories, and understanding. "Wow," one woman said, tears in her eyes. "I feel so seen."

By that point, I had to admit—I did, too. That evening, I took a walk around my neighborhood. As I often did, I exchanged greetings with an older man in the neighborhood, who had long hair and a thick, droopy handlebar mustache. He was usually wearing a plaid shirt, sitting on his porch with a cigarette or a cigar, his pickup truck parked in the driveway. He always responded—emphatically, kindly—and that night I felt a jolt of something: Happiness? Belonging? As I continued to walk, I heard someone practicing the piano up in the hills, the same bar, over and over. It was still beautiful. I saw a lavender bougainvillea lounging on top of a garage I hadn't noticed before. The light was a golden pink as the sun set. *What if I already belonged?* What if you did, too? How would that belonging change our questions—the ones we ask, the way we carry them?

This question was unexpectedly comforting to me. It didn't have to supplant the sense that I wanted more but could instead be its companion, helping me see the soft lines of connection and relationship where I hadn't before.

Soon, I would find other questions, and other people, who would illuminate this idea even more.

Getting Intimate with the Mystery

It was a warm evening in early September, just a couple of weeks after Nearness ended, and I was nervously approaching a building in Oakland. I'd never done anything like what I was about to do before and felt a bit like an impostor. I didn't know any of the rules. (Were there rules? There were probably rules.) And I wasn't sure who else would be there. I tried opening the front door, but it was stuck. Another woman walked up.

"Is it not open?" she asked, clearly a regular.

I tried again, and it swung forward.

Megan Rundel, the host of this gathering, appeared, barefoot, in the hallway, her long wavy hair a mix of brown and gray. "Welcome," she said, smiling.[19] Rundel took me aside: "Do you know what to expect?" she asked me. "Not exactly," I told her. She shared a brief agenda for the evening, and then sent me off into the zendo—a hall where students practice zazen, or sitting meditation. I had an off-and-on meditation practice (emphasis on the "off"), but I was not there for the meditation—or, at least, not only for the meditation.

There were six others sitting on cushions and chairs, two people calling in via Zoom, and a woman rhythmically chiming the meditation bell. She repeated a cycle—a few slow chimes and then several that were sped up, the bells harmonizing with the sounds of sirens and horns outside. Everyone was quiet, staring downward. I looked down at my feet, realizing I'd worn mismatched socks. As others came in, I noticed that they would bow at the door and then bow before their seats—a ritual I realized I'd botched. Ah, well.

What was I doing here? For a decade, I'd intermittently tried to learn more about Zen Buddhism, but I'd never finish the books I started and couldn't quite stick to a regular meditation practice. It all felt too squishy and abstract. But this was a different kind of practice and a different kind of community—another one designed explicitly around helping members become more comfortable with uncertainty.

It all started when I read a passage in a book—just a few short lines about Zen koans. A koan, writes author Estelle Frankel in *The Wisdom of Not Knowing*, is a "short enigmatic tale, a mind-bending paradox, or a simple question designed to take the seeker past the limits of the thinking mind."[20] A classic example of a koan is "What is the sound of one hand clapping?"[21]

"By fully immersing oneself in the koan, body and mind, the seeker becomes the question itself," she writes. We don't find the answer to the koan by thinking. "Rather, its meaning is revealed spontaneously in our day-to-day, lived experience—in the way our actions and attitudes are shaped by its wisdom."[22]

Koan study emerged in China about a thousand years ago, when teachers and students at China's Buddhist Chan school were discovering something strange: Students were having flashes of awakening, or enlightenment to the realities of the world, in the middle of conversations with teachers, fellow students, and even complete strangers. Until then, Chan was introspective, focused on looking inward to find one's true self. But these experiences were challenging that perspective. As evidence accumulated of this different route to awakening, teachers Mazu Daoyi and Shitou Xiqian suggested that your true self is not just inward, but everywhere—everything you see is you.[23] "Awakening, they saw, happens in relationship," writes Joan Sutherland in her book about the practice of koans, *Through Forests of Every Color*.[24] "We meditate together and talk together, we hear birds calling and cars laboring up a hill. As Ma and Shitou did with each other, we find a deep communion with someone we've never met, sometimes with someone who's been long dead."

The teachers and students at Chan also discovered something even more mind-bending. They found that if other people also "brought the story of that conversation into their meditation, they could experience the same thing the participants in the original conversation had: not a lesson or even an understanding about awakening, but awakening itself," writes Sutherland.[25] Koans now include not just snippets of conversation but also poems, folktales, and parts of songs. They are metaphors, in the literal sense: "metaphor" means "to carry across," and koans are like little boats that carry us across our own understanding and experience to new territory.[26]

The koan tradition emerged from a tumultuous time in China's history. In a ten-year period, it is reported that the population decreased by two-thirds. Many people perished from famine, disease, rebellion, and invasion.[27] The instinct of Chan teachers was that to survive in this world, they had to figure out how to turn toward, rather than away, from life. "They had to develop, and quickly, a flexibility of mind, an easy relationship with the unknown, and a robust willingness to engage with life as they found it. . . . For them, Chan practice wasn't about getting free *of* the world; it was about being free *in* the world," Sutherland writes.[28]

Being free in the world had to begin with a very different approach to your relationship with your own thoughts and emotions. Whereas other psychological or spiritual practices are meant to domesticate your mind, koans "encourage you to make an ally of the unpredictability of the mind, and approach your life more as a work of art," writes John Tarrant in his book *Bring Me the Rhinoceros*.[29] Koans do not demand that you believe in any particular religion or that you follow a specific spiritual practice. Quite the contrary. "The surprise they offer is the one that art offers: inside unpredictability you will find not chaos, but beauty. Koans light up a life that may have been dormant in you; they hold out the possibility of transformation even if you are trying to address unclear or apparently insoluble problems."

Koans contain questions, and by studying them, you become connected to the thousands of other people who, across centuries, have asked the same questions and have grappled with similar uncertainties.

Koans, in other words, represent entire communities inside of questions.

And this Oakland meeting was of a community built around the koans—a koan salon. When Rundel started the meeting, she said we would spend the first part of the evening meditating and holding a koan in our minds while we did. Then, after the meditation, we'd discuss what the koan brought up for us.

Rundel read that evening's koan before the meditation began and then repeated it about halfway through:

> *Dizang [a teacher] asked Fayan [a student], "Where are you going from here?"*
> *Fayan said, "I'm on pilgrimage."*
> *"What sort of thing is pilgrimage?"*
> *"I don't know."*
> *"Not knowing is most intimate."*[30]

As I meditated, I reflected on the part that was drawing me in: "I don't know." How funny, even absurd, to respond to a teacher's question at first with such certainty, that you're on a pilgrimage, and then, in the space of a single question, to discover that you don't even really know what a pilgrimage is. To me, it reminded me a bit of the divided self[31]—doing whatever you think you ought to do, without thinking about why or whether it's right for you.

I don't know. I don't know. It became a mantra, one that felt like bricks lifting off my shoulders, like the coiled muscles in my chest relaxing. The more I repeated it, the more I felt relief, joy, and freedom.

Two bells sounded, signaling the end of the seated meditation and the beginning of the walking meditation. In a single line, we traced the circumference of the building at least fifteen times (I lost count), passing a small shelf of Zen books, a zoomed-in photograph of a leaf, a sketch of a laughing Buddha, a painted scroll with Japanese lettering, and a vase of red flowers.

When we returned to the zendo, Rundel served us cups of peppermint tea and chocolate hazelnut biscotti and asked us to share what had surfaced during our meditation.

One woman, in her sixties, a therapist, said it brought up for her the sense that she'd always thought that by this age, she'd be someone who *knows*, who has their life figured out—but she doesn't. She doesn't think she has anything figured out. The koan gave her some comfort, though. Even if we think we know, we don't. And there's beauty in the not knowing.

Another woman, in her late thirties, talked about how the koan brought up the idea of impermanence to her—that knowing feels like a thing we want to grasp and capture and keep, and not knowing is always moving and changing. This, she said, was something she had been thinking about a lot, as she was in the process of ending a romantic relationship. Her eyes began to water. "You think you can know someone, but you never really can," she said quietly. She swallowed, focused on stopping her tears, and looked down.

Rundel's eyes widened in sympathy. "That is so hard," she said, as the rest of the group bowed their heads. We had all been there.

Another man, in his fifties, talked about how he feels pressure in his life to always know—that this made up so much of his identity. He wondered what it would be like to let that pressure go.

One of the Zoom participants, a scientist, said the koan made him think of the lab. He always began an experiment, he said, thinking he knew exactly how it would go, writing the paper in his head. But then something unexpected would happen. And it would be frustrating at first but wonderful in the end, because it opened up new possibilities.

Another woman said she thought only briefly about the koan and then just . . . stopped. She shrugged and didn't apologize—that was simply what had happened, and it was okay.

After the session, members of the group lingered, comforting the woman in the relationship crisis and commiserating with the woman who'd shared that she didn't know what she was doing, even in her late sixties.

When I'd first heard about the idea of a koan salon, I'd anticipated something more akin to an academic conversational environment, a place where people would be jockeying to be the best, to have the most insightful commentary. But that atmosphere of competitiveness wasn't present in the room. No one was trying to be perfect or right; everyone seemed to feel pulled to share what the koan meant to them, in their lives, at this moment. It was a different way of experiencing our collective uncertainties—with support, care, and curiosity, and without any expectation of answers.

The Solution to Loneliness

Megan Rundel was twenty-four when she came across the first koan that would change her life.

> *A student named Qingshui said to Caoshan, "I am alone and destitute. Please help me to become prosperous."*
> *Caoshan said, "Venerable Shui!"*
> *"Yes, sir!"*
> *"You've already drunk three cups of the finest wine in China, and still you say that you haven't moistened your lips."*[32]

The question embedded in this koan is how someone who feels solitary and destitute, who is reaching his hand out for help, could simultaneously have access to the finest things in life and could also be called "venerable." Qingshui is suggesting that he has nothing, and Caoshan is telling him that this is patently false; he is rich. How to reconcile these two distinct perceptions of someone's existence?

This koan stuck with her, Rundel said, because at the time—"young and rattling around"—she was feeling alone and destitute.[33] In a job she hated and lacking a sense of what she wanted to do instead, Rundel felt disconnected—from herself, from other people. She'd been searching for a sense of purpose, meaning, and direction, and so had been trying on various lifestyles: hedonism, asceticism, allegiance to Marxism. "But it all felt performative, and not really connected to something inside of me," she recalled. "It seemed like other people either had it figured out, this thing I was looking for, or they weren't really looking for it. I didn't know how to connect with other people around some of the bigger questions I was asking."

One night, Rundel went to the grocery store and bought a filet mignon and bottle of red wine, thinking, "Maybe this is what I need, something amazing." She didn't have much money but somehow this seemed like an important purchase, a thing that people did who had life figured out, a thing that could help her fill the void.

"I had the steak and had probably most of the bottle of wine and just felt kind of awful," she tells me. "And that sunk me even deeper into despair. . . . It didn't really occur to me at the time that I was having my steak and wine by myself, and that there wasn't actually any meaning in it. I was looking for transformation, a kind of internal, personal transformation, and none of the things I tried were transforming me."

When it came to the koan, Rundel could relate to the experience of a young person approaching a teacher and saying, *I feel alone, please help me.* She was that young person. She had first heard the koan from a Zen teacher during what's known as a dharma talk, a public Buddhist teaching.

What she *couldn't* relate to was the teacher's response—the idea that he, Qingshui, or that she, Megan, had already had the very best of life. "I thought, what the heck?" she recalled. "It felt annoying, like, are you flipping me off? And then I thought, well, this is a koan, so maybe there's something here."

For Rundel, the koan was a kind of emotional sculptor, helping her to shape her diffuse feelings into a series of questions: *Is there help? What does help look like? What do I already have that I'm not recognizing?*

"Forming the question was more important than getting any answer," she said, because the question itself helped her feel less alone. "This is a question some guy had in China in the year 800, and that people have been working on ever since. I could really feel it as a question that a lot of people, not just in my current cohort of young people, but that a lot of young people, people of all ages, had throughout space and time."

This meant that she was "not just an angst-ridden or depressive person. This is part of the human condition." The question, the connections it gave her to countless others, and the conversations it inspired with her Zen teacher didn't cure her loneliness. But seeing how it connected her to others helped her give the feeling more space. Seeing that allowed her to bring more compassion and warmth to the feeling.

And it changed how she perceived the role relationships played in individual growth. As an introvert, Rundel didn't gravitate toward other people to help her sort through personal challenges. But the koan work, especially the conversations it inspired, "opened up something I would never get to individually, if I were just sitting there. That was a complete revelation to me, and it created a whole different paradigm in my mind of what's possible in relationship, in conversation. Something that's so much bigger than the sum of the parts can be created."

This revelation is part of what led her to become a psychoanalyst.

In Session with Rundel

When I first met Rundel, weeks before the September gathering, we were sitting in her bright, plant-filled Oakland office, across from her red psychoanalytic couch, à la Freud.

"People don't realize this, but psychoanalysis is alive and well," she tells me, smiling when I ask about it.[34] "Some people do like to use the couch because it's a nice way to not have to do the social cueing [of looking directly at someone], and it's a way to be in deeper therapy."

For most of her life, even before that moment of feeling destitute in her early twenties, Rundel has been trying to get deeper—with herself, with others. It's part of what led her to begin practicing Zen Buddhism as a nineteen-year-old, though it took her some time to find her place in that community. At the time, she was a sophomore at a liberal arts college and was taking a class on world religions. Though she'd always been curious about spirituality, she grew up in a family of scientists where those discussions were dismissed.

"It was a received wisdom in my house that science is the answer to everything, and that the way to know things is to understand and validate them," she said. It wasn't that she didn't value science, but that she also had "an intuitive feeling that science wasn't the only way to know"—that there must be some value, too, in the arts, in poetry, in literature.

And, perhaps, in spirituality. She was transfixed by Buddhism in particular. Her professor held zazen, the term for seated meditation, before class.[35] She began attending these sessions, and reading every book she could find about the religion.

It wasn't just her classes that sparked her interest in exploring other ways of knowing. That semester, she also took psilocybin mushrooms for the first time. The psychedelic offered a glimpse into the expansiveness of her experience, into a sense of truth about herself and the world. She thought continuing with the Zen work would help her continue to explore that truth.

It did, and it didn't. Though she became part of the Zen community, many of the rituals and conversations felt disconnected from her experience. It was, she recalls, a male-dominated space that could feel unwelcoming to young women.

This changed one summer afternoon, when she attended a workshop led by Joan Sutherland, at the time a senior student at the Pacific Zen Institute. Rundel was drawn to "the clarity, the humanity, the honesty" of Sutherland's speech and thought her gender gave her a distinctive lens. "She brought in real life, the body, the emotions" into her teachings in a way that spoke directly to Rundel's experience.

Perhaps that was because Sutherland, too, had been on her own quest to find a sense of connection with the koan ancestors "in spaces that often felt constricted by a narrow interpretation of the tradition." This was partly due to "men inheriting the tradition from other men and confusing gender norms with ultimate truth," Sutherland recalled.[36] She'd earned her master's in East Asian languages from UCLA and first worked as a feminist antiviolence activist. She was also a book editor for the archaeologist Marija Gimbutas and for other women artists.

Though Sutherland was introduced to koans and Zen as a teenager and could translate them from the original Chinese, she, like Rundel, found herself looking for a way to "bring the koans into our daily lives." She hungered for a community that felt less rigid and more intimate than the others she'd been part of. Whereas koans had traditionally been studied one-on-one, with a teacher and a student, in the 2000s Sutherland pioneered the first koan salons—like the one I had attended in Oakland. These were communities of people who came together to explore the koans, to learn from others' perspectives, and to support one another in carrying the many personal questions that emerged from this practice. The salons took place not in a meditation studio but inside people's homes.

A Community That Accompanies, Rather Than Commands

The salons began with a group of koan students and expanded to their friends—artists and activists who were burning out from their work. Sutherland's intention was to create a set of cultural norms different from the ones these participants—and indeed, most of us—were accustomed to. Inside the salons, during conversations about the koans, "no one was trying to fix or solve anything," she said.[37] People weren't rewarded for pontificating, showing off, or claiming superior knowledge.

In those days, Rundel recalled, "We would talk about koan culture, and it's about not going for the fast and easy answer. It's not parading knowledge or wisdom but embodying deep listening to each other."

A big part of what made the salons work, Rundel said, was Sutherland herself. She exuded warmth and generosity in Zen spaces that, despite their teachings to the contrary, could feel cold and rigid. And in her salons, koans were infused with the messiness of life. "In groups

I practiced with early on, there was more of a feeling of separation, of a kind of special Zen esoteric realm," Rundel describes. "But the one thing that really attracted me to Joan and the communities she has formed is [the emphasis on] being on earth, living into our own questions, linking the koan question to a personal situation. Otherwise, why do we do it? I'm not a philosopher. It feels like it's something to really make use of."

Working with the koans, Rundel said, is "not about giving some Zen answer. It's about, What is this in your life? Okay, so your dog is sick. How is this koan interfacing with that? How might it help you? How might it shift you in how you can hold this piece of suffering or piece of joy? We really try to integrate it so it's not this precious thing outside of real life."

Sutherland never created an official set of rules for the salons, but customs developed out of what people found most helpful. When "people came into culture with more traditional expectations of what a Zen group would be like, or what the practice would be like, they were often disappointed because it wasn't that kind of individual, militaristic, slightly authoritarian style," Sutherland told me. "There wasn't a way for them to gain status and authority based on the perfection of their practice. Someone would speak from that place, try to dominate, have the right answer, sum everything up, and everyone would kind of go 'huh.' People either realized there were no receptor sites for that kind of attitude and tried something else, or they left."

The dynamic that emerged instead, she said, was something richer: "a group of people willing to accompany others through really hard things and really wonderous things."

To *accompany*—not to try to steer the ship of someone else's experience.

Koan salons continue to meet within the Open Source, a network of practice communities led by Sutherland's dharma heirs in the Western United States, including in Arizona, California, Colorado and New Mexico.

As I learned more about Sutherland's legacy, I also discovered something surprising: Sutherland had left her teaching post in the community in 2014, and the reasons were somewhat mysterious to me. I wondered why she'd left. And I wondered what this renowned questions teacher might be able to teach me about how to find and cultivate community in the midst of painful uncertainty. To find out, I knew I'd have to visit her.

The Endless Expanse of Community

To reach Sutherland's house, I turned down an easy-to-miss road off California's famous Highway 1, about three hours north of San Francisco. There was a gate at the entrance of her road that opened to a quiet, winding street. When I arrived, she was waiting outside, waving as her dog, Skye, a Basenji-Chihuahua mix, ran up to greet me.

Sutherland first visited this town with her mother when she was seventeen, staying at a nearby inn.[38] Since that visit, she knew she'd want to live there someday. We walked through her home, where she lives with at least a dozen statues and images of the Buddhist deity of compassion, Guanyin. On her back porch, we had a front-row seat to the ocean, watching the cliffs and rocks come in and out of view through gossamer curtains of fog.

She came to this place because of its spectacular beauty and because she needed to rest. Today, she's in the throes of a serious heart condition, which, like the koans but in its own urgent way, has changed how she relates to uncertainty.

"When you're dealing with a body thing, there's a kind of question inside you all the time," she told me. "When you have an eccentric heart, there's uncertainty about what could happen. You can live in fear of that or you can realize, this is my life. So it becomes a question of 'How do you live welcoming all of life?' And maybe the other side of that is not missing the beauty because of that difficulty."

There is a larger lesson here: Questions and uncertainty aren't like weeds, sprouting in our lives and forcing us to determine how to eliminate them. They *are* life, which itself is full of mysteries. Koans are designed to help people get more comfortable with the mysteries of life by creating more intimacy with questions. Intimacy, in this case, means being as close as you can be to what's really happening to you, without thoughts and feelings altering it.

Intimacy was a concept that at first seemed squishy but ultimately ended up teaching me something important. For me, the confusing part about intimacy was that you couldn't always experience it by trying to understand, by doing more, by asking more questions. Intimacy, I learned, was about simply *being with* someone or something—about accompanying rather than commanding.

In Sutherland's salons, being with someone meant people were not there to give advice or suggestions. People were not there to fix things, solve problems, and provide answers. People were simply *there*—to witness, to reflect, to be moved, sometimes to provide another example from their own experience of the moment being illuminated. But this wasn't a passive or easy exercise; people came to see how "helping" could sometimes be a defense against experiencing the reality of something, while "accompanying" could take everyone into new territory.

Being with something, like an idea, meant that discussions were exercises not in deducing a right answer—*This is what this koan means!*—but rather in layering experiences or placing them side by side, as in a mosaic. The question was, What could emerge when students observed the idea mosaic?

It could be beautiful, Sutherland said, but also messy. And the messiness had flavors. One flavor of messiness nourished the group. This was the flavor Sutherland called "bursting the bliss bubble." During a koan retreat, which was essentially a koan salon extended over days, the first phase would feel both relaxing and exhilarating as people detached from aspects of their everyday lives. "And then always, always something would happen that would bring it crashing down," she recalled—like conflicts between salon members. But this was an essential part of the experience. "You couldn't just create a bliss bubble for everyone to live inside," she said. "You had to break the vessel so something new could emerge. We would work through it all, and then the thing that happened afterward was so much realer than the initial bliss bubble."

What students were experiencing, she suggested, was a shift in their definition of what it meant to be safe and secure. Instead of defining safety as being protected from difficulty, safety meant being able to sit with any kind of difficulty.

People could sit with any kind of difficulty if they knew they were not alone in their suffering. And this was where koans could help—in pushing students to redefine what it meant to be in community. People were part of it, but the koans helped show students that existing in community did not always mean being with people. Koans represented questions that thousands of people had been asking for millennia; studying them was itself a kind of communal connection across generations.

And koans were also designed to clarify how much broader community could be. "It's the crows, and the wind, and the cars driving by," Sutherland said. "The whole point is to be as intimate as possible with all of that, to have that sense of kin. That's not an answer to the questions, but it's the response to living with questions—creating a sense of kinship with as wide a field as possible."

Today, living without other people around in an isolated place, Sutherland told me that she sees "everything alive" as her community. "It's different to see the bishop pines, the pelicans, the mist coming in, as much a part of my community as people," she said, looking out at the expanse.

When I met her, Sutherland was nearly sixty-nine—not old, by any stretch. She'd come to realize that working in community was no longer sustainable. "When you hang out a shingle as a teacher, it turns out there are certain expectations that most students have about your availability to them—in both quality and quantity of time," Sutherland explained. "As my health declined, I could no longer meet those expectations, and we couldn't seem to reimagine the relationship in a way that attended to everyone's needs. So I believed I had to withdraw from a role I could no longer fulfill, and see if there was another way to teach. That turned out to be writing."

Make the Mountains Dance

Our time together was running out, but there was another question I hadn't yet asked Sutherland. It turned out I didn't have to.

"Do you want a koan?" she asked me.

"Yes!" I said excitedly. I told her that there was one I read recently that spoke to me: "Make the mountains dance." The mountains, I told her, have been the site of some of my most transformative experiences.

It was an alpine experience, after all, that had helped A. and me recommit to each other, finally allowing me to let go of the astrologer's disturbing prediction. A few months after our videoconference marriage ceremony (a necessity during COVID-19), we began training to summit Mount Rainier, a 14,410-foot peak in Washington state known for being an arduous, technical, and potentially deadly climb.[39] Roughly half of those who attempt to summit succeed each year;* and as of 2024, more than four hundred people had died on the mountain since the National Park Service began keeping records.[40] As the most glaciated peak in the contiguous United States, climbing it required special gear, like ice axes and crampons. In the months leading up to the climb, A. and I began spending weekends trudging up local mountains, thirty-pound packs strapped to our backs, to prepare for the altitude, terrain, and mercurial weather conditions.

Our training coincided with other life events: I decided to leave my job to work on a big creative project. But after a few months, the idea I had wasn't coalescing, and I was burning through my savings. Depressed that the risk I took seemed to have gotten me nowhere, I felt

---

* The National Park Service stopped capturing this data in 2019, so this estimate is based on prior years.

insecure and worried about my future. But instead of exploring those feelings in myself, I sought constant validation, and attention, from A. He could not give it to me and grew increasingly irritated the more I wanted it. I *hated* being clingy, but here I was—feeling desperate, alone, and unable to get the support and care I felt I needed, in part because I was not giving it to myself first. I was seeking answers from him, in him.

But I didn't realize that then.

All I thought I knew was that it was A. who was the problem—*my marriage* was the problem. He didn't care about me—or he couldn't show me that he cared in the way that I needed him to. Every fight we had, every tense conversation, seemed like more proof that our differences were irreconcilable and that the astrologer was right: We shouldn't be together. I traversed the "Would I be better off with another partner?" path hundreds of times before finally, three days before the climb, I couldn't sleep. The question was keeping me awake because it had become something heavy, hard, and real. Suddenly, it wasn't a boulder tied to my foot, keeping me stuck in place. Instead, it was like a rock on my accelerator pedal, speeding me toward one inevitable action: telling him and ending the fledgling marriage.

"I think we should get a divorce," I told him the morning before we were supposed to drive to Seattle. I explained, tearily, how I'd been feeling—hurt, abandoned, like he didn't care about me anymore. He listened, telling me that he was shocked and sad that I'd been feeling that way for so long and hadn't told him. And he was also angry. I'd led him to the edge of a metaphorical cliff, blindfolded until the last moment, and right before we were supposed to be at the edge of *actual* cliffs.

Should we still do the climb? Could we possibly still enter the focused mental state we would need to summit? These questions felt even more important in light of the especially perilous conditions on the mountain. Just a couple of days before our scheduled ascent, meteorologists reported a record-breaking heat wave scorching the Pacific Northwest. Temperatures in Seattle soared to 104 degrees, and the normally freezing temperatures on Rainier surged to 73 degrees.[41] This heat was dangerous because it melted the glacier, leading to possible rock and ice falls that could impale or kill hikers like us.[42]

The guides assured us they would take every precaution to keep us safe. With so much time and effort invested in the training, we decided we should at least try. And besides, the twelve-hour drive to Washington would give us lots of time to continue our smoldering conversation, which pinballed between mountain logistics, the future of our relationship, and long stretches of tense silence.

After two days of driving, we still didn't have any answers to our relationship questions. But we shifted our focus to the climb as we lugged our bags up to the camp-like hotel. There, we'd stay with the rest of our group before starting the climb. That first afternoon, we met with our guide so he could assess our gear and make sure we were properly prepared.

"All right, everybody, welcome," he said, looking at us sternly. "You know you can die on this climb, right?"

He told us that because of the heat, we needed to adapt the normal ascent schedule. The first part of the climb would stay the same: starting in the morning at Paradise, the base of the mountain, and ascending more than 4,600 feet across 4.5 miles to reach Camp Muir by late afternoon.[43] But instead of starting the summit bid in the early morning hours, we would instead set out around 9:30 p.m., giving us the best chance of avoiding the ice fall that could come once the sun rose.

That first day, we made it to Camp Muir at around 4:00 p.m. The guides instructed us to eat our freeze-dried dinners and then try to sleep in our tiny mountain hut, where we were crammed in with fifteen other hikers.

We did not sleep. Many people in the hut seemed to be suffering from gastrointestinal issues, likely from those freeze-dried meals. A lucky few drifted off right away. A. and I lay awake for hours, listening to a symphony of snoring and flatulence.

Right around the time when we might have been able to fall

asleep, it was time to get up and start the summit bid. We switched on headlamps, strapped on crampons, and grabbed ice axes. At the start, as we headed into a section known as Disappointment Cleaver,* our group was divided into smaller teams, each paired with a guide. Unfortunately, our guide was the least experienced of the lot. This was her first time leading a tour group up the mountain, and she kept getting lost, leading us up steep, rocky paths in the dark and then apologizing as she told us we'd need to double back.

Her flailing guidance exacerbated my own growing concern about whether I could do this. I'd practiced using crampons on glacier, not bare rock, and only in the daytime. Now, shrouded in darkness and tied to the other members of the group with rope, I was being pulled faster than I wanted to go in such unfamiliar conditions. As my exhaustion mounted, each step felt shakier than the last. One wrong move and I could tumble off the ledge.

When we finally reached the first resting checkpoint, I'd never felt so mentally and physically depleted. And this was just the beginning. How could I possibly continue? Looking around me, I saw other members of the group asking themselves the same question. Shocked and glassy-eyed, several decided to turn back.

As I plopped down on a rock and chewed mournfully on an energy bar, I felt tears start to prick at my eyes. What had I been thinking? This was too hard. I was going to hurt myself. A. sat down with me as I breathed heavily, feeling a sense of terror and foreboding, as if to go any farther would be to stumble knowingly into my own demise. For the second time that week, I confessed my doubts to him, my overwhelming desire to turn back.

"Hey," he said gruffly. He was still angry with me, I knew. But he sat down next to me and put his arm around my shoulders. "I know you're struggling right now, I do—but trust me, we trained for this. Every single thing we did for the last four months was to prepare us for this moment.

---

* Interestingly, the origins of this area's name are mysterious, but I found it an apt name for the rough and demoralizing terrain.

We're exactly where we want to be, and we can absolutely do this."

I sighed deeply, letting his words mingle with my terror. We'd prepared for this uncertainty, for the treachery of the unknown, strengthening our capacity to move through it, with it. Still, after the past couple of days, it would have been easy for him to abandon me during this moment of doubt, to shrug his shoulders and let me wander back down with the other group members who were turning back.

But he didn't. In that moment, like a coach in the fourth quarter of a championship game, he shook me out of my self-focused rut and reminded me that we were on the same team. I could rely on him, and he on me. We had trained for this *together*, and *together* we could continue the remaining fourteen-hour-plus journey.

"Okay." I exhaled as I pushed myself up. "Okay. Let's do it."

Early that morning, right around sunrise, we reached the top. At that point, A. was the one who needed encouragement from me; his altitude sickness was worsening. I made sure he was drinking enough water, pulled snacks from his bag, and rubbed his back. We took a few requisite summit pictures, both of us looking half-dead, and then began the long descent.

Later in the afternoon, we collapsed on the pavement in Paradise, where we had started our ascent. Exhausted and exhilarated, we waited for the buses to pick us up and return us to the hotel, and real life.

Of course, this experience did not magically solve all of our relationship issues. Instead, it clarified and revealed their real shapes, like a sculptor chiseling a block of stone. On the mountain, in a space where decisions were life-and-death, we remembered some basic things: that we loved and cared about each other more than anything, even if we weren't always the best at communicating it; that we wanted to make our relationship work, even though one of us (me) might have, for self-protective reasons, suggested otherwise.

In the past, climbing mountains had helped me discover who I became at my limit. This time, it helped me see who A. became, too.

Mount Rainier was also the place where my question morphed into a new one. Instead of "Should we get a divorce?," I began to ask,

"What would have to change for us to stay together?" and "How can we be together?" These were the pawpaw-style questions we would begin to explore, together, over the coming weeks and months.

"The 'make the mountains dance' koan feels really good for you," Sutherland told me, breaking my reverie. "Do you know the story of it?"[44]

The koan comes from the Yamabushi, mountain monks who live in the Dewa Sanzan mountains of Japan, which contain three sacred peaks.[45] Their monastic practice is part meditation and part marathon running through the mountains, Sutherland told me. "The enlightenment of the Yamabushi is when they can see the mountains dance."

She encouraged me to bring the koan into meditation, to carry it with me on a hike, to ask: Are the mountains dancing? What would it mean for the mountains to dance? Though my response to the koan may at first come from external sources, she said eventually it would come from inside me. She gestured to show something like a spout of water emerging from my torso.

Then, she spoke directly to the perfectionism that I, and many people, bring into the relationship with a koan.

"Often the response to a koan is: I have no idea, this is really scary, I'm never going to be able to do this, or what does this have to do with me?" she said. "There's a lot of stuff that comes up. And that's all it. That's your first relationship with a koan. There's a sense of companionship. And awakening happens in relationship. No one just spontaneously combusts and gets enlightened. We're doing it in this vast web of relationship."

Sutherland left me with a final reflection, and the koan that she's currently carrying with her: "The stone woman calls them back from their dream of the world."[46]

What can make the koans so transformative, she said, is that they come from someplace else—not from our own psyches. "They don't take the place of our personal questions, but the two together can be powerful," she told me. "They can shift and change us."

As I left, I felt my koan, like a smooth stone in my pocket. I played around with repeating it to myself as I walked: *Make the mountains dance*. And I wondered how many people, like me, were carrying around the same one—the heavy joy of a mountain in a pirouette, tap dancing, busting a move, getting down.

When Community Is in the Questions

Back in the early 2000s, Oliver found himself in serious trouble. As a freshman in college, Oliver had recently borrowed a book from another university on an inter-library loan. When he returned the book, the student working at the desk asked him to wait for a moment. "The librarian wants to talk to you," she said, turning around to alert her boss that the delinquent had arrived.

The librarian walked out from a back office, a stern look on her face. "If you keep ordering things like this, we'll have to suspend your library privileges," she admonished Oliver at the circulation counter, within earshot of the other student workers. For a lover of books, this was quite a threat—and deeply humiliating. The book Oliver had ordered: *Stone Butch Blues*, a famous novel by Leslie Feinberg about what it was like to be transgender in the 1970s.[47]

"I thought, 'That's a little harsh,' but I didn't order anything else," Oliver acknowledges.[48] Oliver was attending a conservative religious college, one where prospective students had to declare virginity on the application form.

Then came strike two. Oliver, who was assigned female at birth, had been writing his then-girlfriend romantic emails. The girlfriend's roommate read the emails and reported Oliver to the university authorities, who put Oliver on probation during senior year. They also forced Oliver to attend therapy sessions where they tried to convince him that he wasn't queer. "It didn't work," Oliver tells me with a smile.

At this point, Oliver had been holding questions like "Am I gay?" for years. And for many years, he says, "I knew the answer, but I didn't *want* to know the answer."

Oliver grew up in rural Pennsylvania in a conservative religious family of reformed Baptists. That meant that for Oliver and his twin there were lots of rules, dictated by both their family and the church. Men dressed in a certain way; women dressed in a certain way. Men could go to college, women couldn't go to college. Men could choose whom they wanted to marry, and women couldn't. Because secular educators couldn't be trusted to impart the right lessons on impressionable youth, Oliver and his siblings were homeschooled.

"That was a scary world to grow up in," Oliver recalls, particularly for someone who "had zero interest in getting married" and "only liked books." Oliver remembers seeing a queer couple for the first time in the local mall, holding hands as they walked through a corridor. "Everyone turned around and denounced them, loud enough to be heard, but mostly for themselves," Oliver says. "I remember feeling really conflicted in that moment because I felt a certain fascination, and resonance, with these people. But I knew it wasn't okay."

As the countdown to marriage began (in their family, girls were expected to be married by the end of high school), Oliver decided to act, writing a letter to his father, begging him to allow Oliver to attend college at a nearby religious school. Oliver's father considered this letter for a year, until finally, three weeks before school was supposed to open, he gave his permission. Oliver quickly applied and got in.

In college, Oliver's questions of identity became louder.

"I had always had close relationships with women friends, but it was in college where I was like, this is more than that, this is something else," he recalls. "But it was against the student code to be queer."

Soon, Oliver's questions became *heavier*, like bricks in a backpack, weights around his ankles. "What kind of gay am I?" Oliver wondered. "Will anyone love me? Should I come out, or remain in the closet?" Oliver's parents were paying for college, so he was afraid to come out before he graduated for fear that they might cut him off. That meant Oliver didn't have anyone to talk to, aside from his little sister, whom he trusted enough to confide in. Though it was a relief to share who he really was with her, this didn't lessen the load of some of those other questions, which he couldn't pose to her, as a straight cisgender woman.

Disconnected from a real-life queer community, Oliver instead turned to books. He devoured queer biographies, novels, and poetry to try to understand what was happening to him, how to live his life. "Through poetry and fiction I met people who had my questions, too," he says. He understood the feelings they talked about, the way they described feeling in their bodies, their hearts, their minds. "That made [my experience] less lonely," he tells me.

If you're a book lover, maybe you, too, have had the experience of a book comforting you at an uncertain time or the feeling of an author seeming to speak *directly to you*. Communities need not be made up of real-life or embodied people in order to help us move through our big questions.

One night, during a study abroad program at the University of Oxford, Oliver recalls sitting in his room, which had a bay window that faced west. There was a ledge where Oliver could sit with his legs dangling out of the window four stories off the ground, watching the sunset, looking out at the trees. It was there that Oliver realized he needed to be "out" in the world.

"There was something about knowing how many different lives were being lived down there," out in the town below his dangling feet, he says. "[I knew] I could live mine how I wanted to live it. I didn't have to be what I was told I had to be." It was at that point, too, feeling strong and certain in his slow-burn answer, that Oliver had another realization: The weighty questions he'd been carrying had made him stronger—better able to face the uncertainty the next chapter of his life might bring.

It wasn't until graduate school in Chicago that Oliver found a queer community in real life. Landing in the city, "I suddenly saw queers everywhere," he says, reminiscing about his introduction to the vibrant culture. Early on in graduate school, Oliver came out as a trans guy.

What Oliver had been experiencing was technically a form of gender dysphoria, or the uncomfortable and sometimes painful feeling that your gender identity diverges from the sex you were assigned at birth. Researchers estimate that 1.3 million American adults and 25 million people around the world identify as transgender, and 0.5 percent of American adults experience gender dysphoria.[49]

But formal medical descriptions don't always line up with individual experience. Oliver chafes at the term "gender dysphoria," noting that his experience was different, more nuanced, than the clinical definition. He always felt that the official words didn't fit.

"It wasn't so much that I felt uncomfortable in my body," he explains. "It was, 'How do I find my body under all of these expectations people have of it?'"—expectations for how he should be walking, talking, sitting, being.

Treatment for gender dysphoria often includes hormone therapy and later gender confirmation surgery; Oliver chose to do top surgery, removing his breasts, a year after graduate school when he had a job that would pay for it. He decided against hormone therapy, which he acknowledges would make it easier for him to "pass" as a man in certain contexts, because "I also like being in-between. I like being readable as queer—not just in sexuality but in gender, and my way in the world." He wanted his protest against the expectations of his body—how it's supposed to act, whom it's supposed to love, what it's supposed to look like—to be visible to all.

Though the queer community in Chicago was welcoming of him, none of Oliver's new friends shared his trans identity. So, as he had done before, Oliver got creative about how he would find that community to support him, as he lived into more questions. This was the mid-2000s, so he turned to YouTube, still a novel platform, where there were an increasing number of videos featuring trans people. Before long, he had developed a daily ritual: Wake up, make coffee, watch YouTube videos. "It was a way for me to say [to myself], 'There are a lot of people like you. You make sense. You can handle whatever comes today. Other people are handling things day by day, too.'"

Holding onto questions for longer, allowing them to ripen or turn into new questions, has made him "more willing to accept answers and try to live paths that are not as clean-cut, not as obvious or common. There's something about living with what feel like risky questions that gives you sea legs for the higher-risk choices you actually take."

Here, Oliver points to a counterintuitive dimension of this practice. "You might think of questions as things that cause you to lose your footing, that are disorienting," he says. "But actually, they are more like anchors, helping you to get centered in who you are and the journey you're on."

You're Already in Community

It can be easy to have an image of the "right" community to help us love our questions. I certainly did. But communities that support us in uncertainty can look and feel very different from what we might expect. Communities don't need to be made up of close friends; the people who support us can even be semi-strangers. In fact, they don't need to be people at all. We can find community in nature, in stories, and within ancient traditions, as well as inside the questions themselves.

What I learned from Oliver, Rundel, Sutherland, and others was that new communities to support me in my uncertainty were out there and accessible. But I didn't necessarily need to feel pressured to find a *new* community, because the truth was that I was *already* in community. I lived my life in a network of relationships. I was connected, even if I didn't always feel that way. And those connections could help me feel more secure and sustained when I felt the most lost.

Though I was beginning to discover a basic, universal foundation to being able to love our questions—a strong relationship with self, with others, and with larger communities—I was also hearing about another, more complicated layer from many of the people I spoke with.

Though all of us live with uncertainty, some of us deal with different intensities and degrees because of the identities we hold, whether they're related to socioeconomic status, race, sexuality, gender, immigration status, or physical abilities. As a trans guy, Oliver, for instance, had to grapple with uncertainty that someone who is cisgender does not. Holding other marginalized or underrepresented identities correlates with all kinds of other uncertainties that those who are part of the majority aren't forced to live with: Some worry that a police officer could pull them over for no reason and that they could get shot because of their race. Others question whether they'll be able to eat this month and pay their bills, whether they'll get deported tomorrow or be able to live in the same country with their family. Many wonder whether people will see them for what they've accomplished, rather than what they look like. Others don't know whether they'll be able to physically access all of the places where they need to be in a day.

On the surface, there was a simple narrative here: Identity is linked to power. The sense of having power, or not having it, shapes the experience of uncertainty, because less power can create the conditions for insecurity and more unknowns. But this narrative, I found, was only the beginning of the story of how identity interacts with uncertainty—a story that contains a counterintuitive lesson about living with questions meant for all of us.

6
Converting Questions into Freedom

THE BIG IDEA

In times of uncertainty, our instinct is to shrink. We think that clinging only to what we know will keep us safe. But many who have been able to thrive in conditions of extreme uncertainty have done the opposite. They've used their questions as tools to find freedom and space in the unknown, redefining themselves, their relationships, and their futures.

In some ways, Rilke lived two lives: out in the world, he was anxious, full of doubts, seeking answers in new people, new places. "Much of his . . . personal experience was desperate," writes Lesley Chamberlain in her book Rilke.[1] *"No conventional religious consolation came his way for feeling lonely and displaced. The irony is the consolation he has since given to so many readers in so many countries. It's an invitation at once to expand inwardly by way of imagination, and therefore forever to enlarge on a world outside us. Rilke asks us to concentrate on inner resources to be able to find, at all, what is out there."*[2]

Inside the self Rilke curated in letters, in the ideas he poured into his poems, he was someone else: a grounded sage who urged not just contentment with life's uncertainties, but an almost worshipful adoration of them.

"Fear of the inexplicable has done more than impoverish the life of the individual, it has also constricted relations between people, lifting that relationship out of the riverbed of infinite possibilities, as it were, onto an unseeded plot of land on the shore where nothing is accomplished," he wrote to Kappus in 1904. "If we think of this existence as a room, larger or smaller, then clearly most people explore only a corner of this room, a seat by the window, a strip of floor on which they pace back and forth. That gives them a certain security."[3] And yet, he continues, this security is a false one—preventing us from becoming more intimately acquainted with all of life, from experiencing the surprises it might hold.

In his writing, Rilke was determined to "cherish life in direct confrontation with what robs us of life," write translators Anita Barrows and Joanna Macy in their book In Praise of Mortality, a collection of Rilke's poetry.[4] His courage as a poet is "born of the ever-unexpected discovery that acceptance of mortality yields an expansion of being. In naming what is doomed to disappear, naming the way it keeps streaming through our hands, we can hear the song that streaming makes."

In 1997, Barbara Schoen knew who she was: a successful management consultant who traveled sometimes every week for work. She was an avid cyclist, biking fifteen or twenty miles a day and toting her bike with her when she traveled across the country. She pedaled in the Adirondacks, around the Fox River near Chicago, along abandoned train tracks and down the Huron River trails in Ann Arbor, Michigan. She was a newlywed, married to her husband for a year. They had a lake house and a pontoon boat where she would sometimes take her laptop out and work.

Schoen was someone who, when leaving the house, made sure she'd perfectly applied her eyeliner and that her handbag matched her shoes. She was the type of person who regularly painted her nails.

Schoen was someone who was going places, even if she wasn't exactly sure where those places were. Though she felt herself being propelled forward, with vague ambitions to move up, up, up, "I never felt like I was doing what I needed to be doing with my life," she recalls. "I didn't feel like I was making the mark I needed to be making."⁵ But who had time to think about that when there was so much to do?

Schoen once told an airplane seatmate that she "wanted to change the world." For her, this meant making educational opportunities available for kids who lacked access to them, just as she had. She grew up in a blue-collar, "barely middle-class" family in Michigan. Her father worked three jobs, including at a factory and an upholstery store, and her mother was a school crossing guard and library technician. There was a lot of love to go around, Schoen recalls, even if her parents were rarely there physically for the five siblings. "We grew up fiercely independent, and fending for ourselves," she says.

Money was tight, and college was out of the question. But Schoen lucked out: After high school, she got an entry-level job working for a Fortune 500 engineering firm. There, she stepped into a world of privilege and what to her felt like glamorous work, traveling to places like Chicago and Toronto. The company paid for her college degree.

Even with the degree and years of work experience, she'd never forgotten the role that luck played in her life. She wanted to help other kids who were growing up the way she had get opportunities to thrive. Someday.

The year before, 1996, both of her parents had passed away within the space of five months—her mother from cancer and her father, she suspects, because he couldn't stand to live without his wife. Then, in 1997, when Schoen was thirty-three, her grandmother died. From her home base in White Lake, Michigan, on the outskirts of Detroit, Schoen traveled to the northern part of the state with her siblings for the funeral. On the way home, her family drove back together in the middle of a snowstorm. Her sister was driving, her brother was in the front seat, and Schoen and her other sister sat in the back. An "avid safety belt wearer" she would even risk ruining a nail to make sure she was buckled in. But the seat belt in this car was choking her, so she spent the trip continually unbuckling and rebuckling. At some point, hungry for a snack, she unbuckled the belt and leaned into the back to get some SnackWell's cookies. When she returned to her seat, she didn't rebuckle her seat belt.

Suddenly, the car started skidding. "Help me, Jason!" Schoen's sister, who was driving, shrieked to her brother in the passenger's seat. Jason grabbed the wheel in an attempt to correct the car, but it continued to slide, hitting a bank toward the side of the highway. The car flipped over six times.

Schoen was the only one who was ejected from the vehicle; to this day, no one, including the rescue team, is sure how she survived. Lying on the side of the road, she heard her sister say, "Where's Barbara?"

"And I said, 'I'm out here. I don't think I'm going to make it.' I could feel myself dying. I said, 'I'm okay, just let everyone know I love them,'" she recalls.

But then serendipity intervened. On that frozen road, the people in the car behind them happened to be a retired paramedic firefighter and his wife, a nurse. They kept her alive until an ambulance arrived six minutes later.

Schoen remembers making a choice at that moment, with the understanding that not only was she seriously injured but she couldn't move at all. One option was that she could die right there. "I wasn't sure I wanted to live like this," she says. "I was very active and athletic." On the other hand, she thought, "If I die now, then it's done. I can always die later if I don't like living like this. So I chose life. They got me in the ambulance and I remember they were panicking because my blood pressure was diving, but they managed to keep me going."

For three days, doctors kept Schoen intubated. They told her husband what had happened: Her spinal cord, with the membrane inside it, was like a tube of toothpaste that had been squeezed out when her neck broke. The doctors had done their best to push it back in. "She'll be lucky if she can shrug her shoulders," the doctors told him.

As soon as they removed her breathing tube, "I said, 'Get me to rehab, let's keep going,'" she says. "I was ready to go and get back with life."

What her new life would look like, Schoen didn't know.*

From Paralysis to Freedom

She did know that surviving the accident would allow her to explore the questions that had been percolating even before it happened: What did she want to do with her life? Who did she want to be? What would "changing the world" look like for her? Slowly, Schoen was able to regain some use of her arms and upper body, but she was paralyzed from the waist down.

She began to look around at the other injured people in the hospital, recognizing that, once again, she'd gotten very lucky: Her auto insurance policy had a catastrophic injury benefit that provided unlimited care for her medical needs. Other folks who'd been injured in non–motor vehicle accidents "had next to nothing. Social workers were searching for someone to build a ramp on a patient's home so they could be discharged from the hospital. Another prominent businessman's wife was being trained to provide daily care for him. This injustice festered in me."[6]

Extreme uncertainty, Schoen learned, demanded that she become more flexible in her identity, stretching into a new one like an unfamiliar yoga pose. Schoen's big questions—some more of the peach or pawpaw variety (Would she regain her ability to move? To live an independent life?) and others more like the heartwood (Who was she now that she'd had the accident? Who could she become?)—along with her changed circumstances, meant she could no longer cling to a past version of herself, who no longer existed. The injury, she says,

---

* One of her sisters, seated in the backseat, suffered no injuries from the impact. The two other people in the car (her brother and other sister) suffered temporary blindness and injuries to their shoulder blades. Despite living across the country from each other, Schoen's sisters have continued to be involved in her life.

"was kind of a relief. I had felt trapped in my current position. . . . I felt like the injury gave me an opportunity to do something else." She decided to return to school and, encouraged by others' fierce urging, to earn her PhD and become a rehabilitation counselor—a person who provides modifications and accommodations to help people with disabilities find careers and maximize their independence.

There were more changes. After her accident, Schoen started leaving the house without her eyeliner. She no longer cared whether her shoes and her purse matched. She learned how to ask people for help. Adjusting to a physical disability, she says, paraphrasing the psychologist Beatrice Wright,[7] meant redefining what was important to her. "It's about renegotiating your goals, and coming up with different questions"—questions that can help you redefine, or even rediscover, who you are.

"What do we have to gain from redefining ourselves?" I ask Schoen, speaking to her on a video call twenty-six years after her accident.

"The word that pops into my head is 'freedom,'" she responds—freedom from the rigidity of our perceptions and self-imposed limitations.

But there was another prison Schoen needed to break free from—one she wouldn't even begin to see until much later.

Identity and power transform our experience of uncertainty, especially for those who hold historically disenfranchised, marginalized, and unrecognized identities[8]—for instance, people who live, as Schoen does, with a disability. For these groups, uncertainty has a different flavor.

What I learned from Schoen, and the many others I spoke to who hold historically marginalized identities, is that power and identity transform our experience of uncertainty, of living our questions—but not always in the ways we might expect. Power and identity enrich and impoverish, create and destroy. Schoen found she became both more limited and more free at the same time. Uncertainty, questions, and identity interact in complicated ways.

What, I wondered, could the experiences of people who deal with greater levels of uncertainty teach all of us about how to live better with it?

Using Questions to Expand Possibilities

When he was a kid, Miguel thought he knew how to cope with the storm clouds of uncertainty hovering nearby. His coping mechanism? Education. Doing well in school, he figured, would be his life raft into a sunnier and more certain existence, one with fewer questions and more answers. With good grades and a college degree, "everything would work itself out."[9]

Born in Guasave, a city in the Mexican state of Sinaloa, Miguel was two when his parents migrated to the US. They came for better economic opportunity.

Miguel spent his middle school years with predominantly Black and brown kids in a lower-income neighborhood in Phoenix, Arizona. There, he began to realize that as an undocumented individual in America, he would have to "work harder in order to be successful," to "jump through hoops." If he wanted any chance at rising above his undocumented identity, he'd have to excel in his classes. Even if school let out early, he would head to the library, which was open after hours.

"I had to mature very young," he said, recalling how, even when he was spending time with friends, his mind would constantly wander to everything that could go wrong—jeopardizing his status in the country. "I couldn't just enjoy being a kid."

Miguel's hard work paid off, earning him a spot at a private preparatory high school in 2014. There, he discovered a different challenge: belonging. For the first time, he felt lonely. Most of the other students were white. They went on vacations to other countries with their families; his family couldn't leave the US. They could get driving permits, and then licenses, when they turned fifteen; Miguel couldn't. He also felt he couldn't share his immigration status for fear of how it would be received.

When Miguel was sixteen years old, in 2016, he became a recipient of DACA—the Deferred Action for Childhood Arrival. It's a policy that allowed certain undocumented immigrants who came to the US as children to apply for eligibility for work permits.[10] It is a two-year shield against deportation, which costs more than $500 to renew every two years. But since the policy was implemented, it has become highly politicized, its power dependent on who holds the presidency. Meanwhile, federal courts have issued mixed rulings on its constitutionality.[11] DACA is "always in the news: Will it end this year? Will it keep going? We're in this constant limbo," Miguel, now in his early twenties, told me.

So as political battles roiled, Miguel focused on school, continuing to grind away, choosing to focus on getting the right answers to his homework rather than ruminating on the murky uncertainty of his undocumented status and what it would mean if he couldn't stay. Education would allow him to prove his value, cement his worthiness as a resident of the US. With a college degree, he thought, it wouldn't matter as much that he was undocumented or a DACA recipient.

What he wanted to study was also becoming clearer. It wasn't too often that a high school student elected to spend their free time doing *other* friends' homework, but Miguel found he was insatiably curious about the psychology coursework of his buddies. He couldn't take the class because of his schedule, but "we would have a break period every day, and they would be doing their homework, and I would join in with them to try to answer the questions." He was fascinated by how the brain works and why people do the things they do.

After graduating from high school, Miguel got a scholarship to a university where he continued to study psychology. For the first couple of years, he worked with the same intensity that he had in high school. He was there to earn a degree, nothing else. But then, something changed. He became friends with students who shifted his focus, illuminating an issue that he couldn't unsee: Their university served a majority of Hispanic students and classified as a Hispanic-serving institution (HSI), meaning it was qualified to receive aid from the government.[12] And yet, the school administrators and professors were predominantly white.

"When we talked to people in positions of power, [the issue] was brushed off and put back on us," he said. He, along with other students and a professor, began to question those leaders.

Miguel was starting to learn an unexpected lesson: that maybe there were other reasons he'd spent his whole life questioning his worth, feeling the need to prove it with educational achievement, and wondering why he was never treated the way he thought he should be. The problem wasn't *who* he was but *where* he was: a place with certain kinds of people in power—people who hadn't experienced life as he did and yet were making decisions that could disproportionately affect his existence. The problem, he began to see, was not just individual, but also structural.

Miguel's experience is all too common in the broader undocumented community. "What I always remember is hearing people say that people like me should 'earn' our citizenship," writes the journalist Jose Antonio Vargas in his memoir, *Dear America*.[13] Vargas, who was born in the Philippines and came to the US when he was twelve, first came out as undocumented immigrant in a 2011 *New York Times Magazine* article.[14]

Underpinning this sense of needing to prove one's worth are many myths about undocumented immigrants in the US. One is that they are economic freeloaders, taking more than they contribute. The fiscal reality doesn't reflect this: For instance, undocumented workers pay billions of dollars in taxes and Social Security contributions every year.[15] In 2022, undocumented immigrants paid $96.7 billion in local, state, and federal taxes.[16] "Immigrants are seen as mere labor, our physical bodies judged by perceptions of what we contribute or what we take," Vargas writes. "Our existence is as broadly criminalized as it is commodified."[17]

―――

During his junior year, Miguel began to take classes with a psychology professor who "taught very differently," encouraging his students to ask questions and think for themselves. His classroom "was a place of uncertainty," Miguel recalled.[18] Though he was familiar with the experience of uncertainty, he hadn't yet learned the art of asking questions—of himself, of the world around him. This was his introduction.

"I started to ask, what am I doing after this?" he said. He channeled those questions and emotions into art. Though he'd been sketching since high school, he began drawing and journaling "with purpose," an attempt to express his emotions rather than dismissing them.

Through this introspection, he realized he didn't want to go to graduate school. As a first-generation college student, he felt pressure to get a job and start making money. But *what* did he want to do? Wasn't his undergraduate degree supposed to make all of that clear?

"My parents had a certain expectation of what they wanted me to be and what they thought I should be doing," which, he said, included finishing his studies and earning a degree in a field that would "guarantee access to a higher socioeconomic status." That way, he could support and care for them. "I was also trying to deal with [those expectations]: Should I do what they want me to do? Should I do my own thing? They've done so much for me, so do I have some responsibility to fulfill?"

How could he find a job if he still didn't know not just *what* but *who* he wanted to be?

After he graduated in 2022, he returned home to what used to be his summer job: helping his dad paint houses. He kept wondering, What was the point of his education if he was just going to do the same thing?

He started to grow more anxious. And at the same time, he had a realization. He was finished with school, but the questions in his life weren't going away. Education, he saw, had served as a tool to forestall the discomfort of uncertainty, to make him feel valued in a world that tried to tell him otherwise. But now, he no longer had school to dictate his next steps. *What if I can't make it in the US?* he wondered. *What if I can't get a job?*

Those existential questions were tumbling around in his head when he took a trip to Mexico in January, nearly eight months after he graduated. Known as an advance parole trip, it allows DACA recipients to go to a different country for humanitarian, educational, or familial purposes and return to the US.[19] The purpose of this trip, which he took with a group of other DACA recipients of all ages, backgrounds, and occupations, was to learn about the culture of Oaxaca, Mexico. But for Miguel, the experience held another lesson.

On the second night of the trip, the travelers crammed into a hotel room to hang out and have drinks. As they stood in a circle, a married couple in their late twenties suggested that everyone go around the room and say where they were from, what they did for a living, and anything else they wanted to share.

Like Miguel, the couple was from Phoenix. But one of them was adopted in addition to being a DACA recipient, and he'd spent his childhood moving from home to home. What struck Miguel most was how differently this man had dealt with uncertainty. Instead of allowing the uncertainty to constrain his possibilities, it multiplied them. "He took his experiences and decided that they would not define him," Miguel recalled.

Miguel listened to story after story from the group, each unveiling a different approach to a similar root experience of being undocumented. One person was a backpacker. Another worked as a mortician. The experience, he said, was transformative, prompting him to question how he had been approaching his life. He began to see that he'd been making decisions from a place of fear. The uncertainty in his life had kept him tethered to a narrow path, preventing him from doing anything too spontaneous or risky. But hearing the stories of the group "showed me there isn't a set way in how I can do things. The fact that I'm a DACA recipient shouldn't box me in. It was enlightening to see that you don't have to live in such a strict way in order to be happy, even if you're undocumented."

It also made him see that, for much of his life, he'd been silencing the questions that whispered inside of him—the kinds of questions that might have helped him get back to what he really wanted, separate from the expectations of others or how he imagined others might see him.

"Since I graduated, I'm [now] not just pressing the off button on the questions," said Miguel, who currently works at a nonprofit that supports individuals who are undocumented or DACA recipients, as well as mixed-immigration-status families. Instead, he's facing the questions and asking, "What other things can I do even though there's this big question of uncertainty? It's taking those big unanswerable questions and seeing what answerable and positive questions come out of them. I don't know what's going to happen, so I might as well do something I want to be doing. Instead of uncertainty controlling me, I can control what I choose to do." Instead of allowing uncertainty to keep him stuck in an anxious state, he would use it, and his questions, as tools to move him forward, to explore other possibilities for his life.

How Uncertainty Expands and Deepens Our Relationships

On a chilly night the week before Thanksgiving in 2005, a group of fraternity members in New England carried a pledge up a mountain in the rough terrain of the northern stretch of the Appalachians. The members wielded flashlights as they deftly moved the man on the green canvas stretcher over boulders, logs, and streams. Leaves crunched underneath their feet; the wind shook the trees, bit their cheeks. The conversation consisted mostly of logistics—"Do you have him?" "I have him." "Can I pass him to you?" If one person tripped, the whole operation would go down.

This was not the result of a fraternity hazing ritual gone horribly wrong. Quite the opposite. The man on the stretcher was Joseph Stramondo, a college student and fraternity recruit. As a self-identified little person who also has mobility disabilities, Stramondo uses a wheelchair. A few weeks earlier, Stramondo and one of his new friends, Sean, had started scheming about how they could get Stramondo up the mountain for the fraternity's initiation ceremony. The stretcher had been Sean's idea, and he got the other fraternity members on board.

"There's no way I would do that now, but at that point I was surprisingly calm," Stramondo says, recalling the night. "It was fun—an adventure."[20] The men carrying him didn't make him feel like a burden, in part because the experience of carrying him exemplified what they were all hiking up the mountain to do: to learn to trust and depend on one another—to learn about the power they generated from that interdependence.

Much of the uncertainty that people with disabilities face comes from the possibility of "misfit," a term popularized by the disability researcher Rosemarie Garland-Thomson.[21] Misfit happens when someone's body doesn't fit with the environment they occupy, and tension results. For someone with physical disabilities, this often means grappling with questions of access that are built into everyday experiences. Let's say you want to move into a new apartment. How will you find a place that's affordable and also wheelchair accessible? If the place you want to move to isn't already accessible, is the landlord willing to make accommodations? Suddenly, you're dealing both with structures in the world and the attitudes of people with the power to change them. For someone with an intellectual disability like autism, a misfit can exist in noisy social situations—like bars, malls, concerts—and many individuals with autism can become overstimulated in those environments, Stramondo suggests.

The misfit experience is not exclusive to disability. "There are all sorts of ways that people can misfit in terms of gender identity and race—all sorts of ways the world is constructed for a certain kind of human being, and folks who don't fit that mold experience it very differently," says Stramondo, who is an associate professor of philosophy and humanities and director of the Institute for Ethics and Public Affairs at San Diego State University.

For instance, Stramondo's friend Sean also had a misfit experience in college. He was struggling with his coursework. So Stramondo occasionally helped him proofread and revise his college essays. In this way, the misfit experience inspired a distinctive coping strategy for uncertainty—one that Stramondo's fraternity also implicitly understood.

To help explain that strategy, Stramondo tells me about the philosophical concept of relational autonomy.[22] This term might sound funny, because it combines two ideas that are seemingly at odds with each other: "autonomy," highlighting our independence, and "relational," highlighting our interdependence. Relational autonomy argues that we're not all purely independent actors, making decisions in a vacuum.[23] Instead, our decisions and behaviors are influenced by the people and world around us. Seems obvious, but it's easy to overestimate how much we believe we're controlling our own lives.

"What our relationships look like, and how supportive they are, can largely determine for disabled people how we respond to uncertainty" and how safe disabled people might feel, he tells me. "Uncertainty can deepen our relationships and create new relationships with folks that identify with you or that care about you in a way that is not grounded in pity." Relationships, he says, can either support our journey toward integrity, or they can impede it.

Though we're *all* interdependent, people with disabilities tend to be more attuned to these relational dynamics, partly because of the stigma they experience.[24] Sometimes, Stramondo will play a game with an audience when he's giving a talk: Are you disabled or are you a rock star? If you're a rock star and you need help to cook all of your meals, that's called a personal chef. Need someone to drive you everywhere? That's your chauffeur, or personal driver. Need someone to do your grocery shopping? You get a personal shopper.

Now, he says, compare those needs and associations on the part of the rock star with those of someone who has a disability. A disabled person might receive the same kind of care, but "you don't feel like a rock star when you have someone come to help you clean your home when you're a disabled person," he says. In one case, the help is perceived as a sign of power. In the other case, it's perceived as shameful—a symbol of weakness. "Lots of people have housekeepers, but we call housekeepers for disabled people caregivers," Stramondo points out. "This is the way that it is stigmatized, and makes folks aware of the interdependency that they're experiencing."

This awareness extends into other relationships. And for Stramondo, it also means recognizing where other parts of his identity as a cisgender, heterosexual white man might make him less aware of his impact on others. "It would be very easy for me to [say to my spouse], 'Okay, you do all the laundry because I'm disabled,'" he says, emphasizing the research showing that women still do the vast majority of care work and cognitive labor at home—often labor that is invisible to their partners.[25] Stramondo and his partner are careful to pay attention to the balance of chores and childcare between them, knowing that if their dance of interdependence gets off kilter, "it can be exploitative."*

How does he, and how can any of us, know whether our relationships are off kilter in this way? "A lot of it has to do with awareness of how you're feeling," he says. "It comes down to emotion. Are you resentful of people you're in relationship with? Are you shameful about the help you need? Those are the two major emotions to look out for, that might clue you in to something that's off."

For Stramondo, the goal is to find a balance between asking for help and support and considering how to give that help and support back to others. It's about fostering the kind of relationships that allow for a natural flow of give-and-take. This, he says, is a strategy anyone could use to evaluate and invest in relationships, but one that's more often practiced, by necessity, among disabled people.

For example, a professor who had just received an offer of employment from San Diego State University called Stramondo out of the blue, asking him what she needed to know about accessibility on campus. "We became best friends," Stramondo says. "We hang out all the time and help each other deal with the uncertainty that comes from having 'misfit' experiences on campus."

Stramondo and his friend had one of those experiences one day when he was giving a talk about disability on a side of campus far from where his friend normally worked. The talk was for people new to Stramondo's scholarship, and it elucidated basic insights about living with disabilities. His friend attended the talk and showed him extra

---

* After our original conversation, Stramondo and his partner decided to get a divorce; nevertheless, he emphasizes that this balance of labor is still an important one in their evolving relationship.

support by helping to explain key ideas to the audience during the Q&A. When the talk ended and it was time to leave, Stramondo and his friend looked outside: it was pouring rain. This presented a problem: His friend used a manual wheelchair and had to get all the way to the other side of a hilly campus. On her own, she would get soaked—especially because she was unfamiliar with the accessibility ramps and elevators in that area. Stramondo uses a battery-powered chair, meaning he can go fast. But he knew that if he stayed at her pace, his chair would short out in the rain.

"So I looked over and was like, 'How do you feel about me towing you?'" he recalled. Like his fraternity brothers carrying him up the mountain, he was now paying it forward. The two sped along campus, Stramondo's friend holding on to his chair so he could deposit her where she needed to go. This was relational autonomy at work—Stramondo giving support to his friend who had just supported him, the two of them deepening their own relationship, expanding the possibilities for how they could be in an uncertain world.

Uncertainty Shows Us Who We Are

After her injury in that car accident, Barbara Schoen channeled her ambitions toward new goals. Once she secured her master's and PhD in 2010, she began working at the University of Texas as an associate professor, becoming director of the rehabilitation counseling program there and starting to work toward tenure. She was in charge of recruitment, solving faculty and student challenges and hiring. She developed presentations, wrote articles and grants, and collaborated closely with the dean and disabled community. Suddenly, Schoen found herself working eighty hours a week. She would wake up in the middle of the night and ask her nurse to write down the to-do list items spinning around in her head.

Though she'd changed since her injury, one part of her had remained the same. It was the part that asked, "Am I enough? Will I ever be enough?"

"I always felt 'less than,' growing up blue-collar and then working among engineers and other professionals," she tells me.[26] "I used words like 'tooken,' instead of 'taken,' and someone would have to correct me. A mentor told me that I needed to shine my shoes. I didn't know about these social cues, which made me feel ashamed but also gave me the drive to learn more."

She continued to push herself harder and harder, fueled now not by the need to fit in at a corporate workplace but by feelings of guilt about the support she was receiving from her insurance company when so many others weren't so lucky. She became preoccupied with the question whether she was doing enough to give back to others. *Will I ever repay the debt?* she wondered.

"As one of my very wise therapists explained, 'Injury doesn't change who you are—it amplifies it,'" Schoen says. The question of who you are becomes so loud that you can no longer ignore it. For her, this point came when the dean of her school offered Schoen the associate dean position. How, she wondered, could she possibly work more? She couldn't and turned it down. She was beginning to realize that "climbing higher wasn't necessarily better." This, she says, was "a bitter pill to swallow, as I had lived my life climbing." Maybe, she considered, progress and growth didn't always have to mean more, higher, harder. "This whole idea of 'being enough' is artificially manufactured by the world around us"—and it's an idea that's often equated with more, more, more.

Even after this realization, Schoen still struggles to find balance, to nourish the part of her that wants to give back while not fueling the part that thinks she'll never be worthy of what she's been given.* Some days, she says, are better than others. Regardless, she looks for-

---

* For Schoen, giving back has included prioritizing projects to help empower and maximize the independence of people with disabilities. For instance, she received a grant from the Christopher Reeve Foundation to survey the needs of consumers in Michigan and provide a guide of services available for people with disabilities. Through another grant, she designed the infrastructure and secured community partners for an assessment of unmet needs of people with disabilities in the Rio Grande Valley of Texas.

ward to the final chapter of what her life will bring; she is armed with a lifetime of answers, but even more questions, as she optimistically strives for that elusive balance she seeks.

In Schoen's story, I heard echoes of the Pulitzer Prize–winning author Robert Caro's conclusion from his career studying some of the twentieth century's most powerful political actors. "There's an old saying: All power corrupts, and absolute power corrupts absolutely," he told a journalist for *Esquire* in 2009.[27] "The more I've learned, the less I believe it. Power doesn't always corrupt. *What power always does is reveal.*"

As uncertainty amplifies who we are, it can also reveal parts of us that before we'd been able to ignore, or suppress, or deny. At least for me, a big part of what's uncomfortable about living with uncertainty results from resisting this revelation and clinging to an old identity or way of being. When I've sat with questions like "How will I survive this?," "What's going to happen?," "What if I lose what I've worked for?," I'm subconsciously answering them based on how the *current* version of me will survive, respond, or react. In other words, I'm limiting future possibilities based on the person I am now and the ways I've behaved before, instead of who I could become.

To imagine being another type of person, or behaving in a different way in the face of challenging circumstances, can feel terrifying because you feel as though you might lose something in the process. In fact, you may be gaining much more.

I'm reminded of the words of Rilke translators Anita Barrows and Joanna Macy: "In naming what is doomed to disappear, naming the way it keeps streaming through our hands, we can hear the song that the streaming makes."[28] When we acknowledge what we are afraid to lose and our compulsion to grip tightly to what's known, we can start to open ourselves up to what lies beyond it. We can hear the sound of something beautiful, ephemeral, streaming through our fingers, when before our experience was limited to the fact of it slipping away.

This was the challenge of Schoen, Miguel, and Stramondo—to hear the sound that the streaming makes. They all figured out how to take uncertain experiences that threatened to narrow their lives and flip them into an expansion of who they were and who they could be. They saw how cultural narratives sought to define them based on their identities, to assign them worth and strength, to diminish and denigrate their existence. And they wondered how, instead, they could let the uncertainty and questions help them grow bigger than any single definition, to trample over expectations.

I learned from them that uncertainty forces us to renegotiate the relationship we have with ourselves and to imagine another way to be in the world. It's a tall order, but one we don't need to embark on alone—or without a set of tools to guide us.

BRIDGE TO PART THREE

Making Your Questions Map

Renegotiating the relationship we have with ourselves requires asking big heartwood questions like: "Who am I?" "Who do I want to become?" "How do I get there?" Though these stay with us throughout our lives, the key to living them, and loving them, is the key to any new skill: practice. The more we practice asking and paying attention to our questions, the more we can see answers pop up in our lives, and the more we might find they open up into new and better questions.

But how do you practice something that feels fuzzy and uncertain to begin with? Where do you begin when all you feel is lost? You do what explorers have been doing for millennia. You create a map.

Exploring Uncertainty

"Okay, let me explain this by telling you about a very old movie," Nick Kabrél said. Kabrél is a doctoral researcher in psychology at the University of Zurich in Switzerland in his early twenties, and I was talking to him on video chat.[1] "The movie is called *The Lion King*—I don't know if you've heard of it."

"I've heard of it." I tried not to laugh, my amusement mixed with a touch of horror. It was the same thing I felt when I noted that *Friends* was now playing on Nick at Nite—in the same time slot that once played reruns of shows like *I Dream of Jeannie* when I was a little girl. I was getting old.

Kabrél reminded me of one of the first scenes in the movie, when young cub Simba and his father, the king, gaze over the savanna.

"Look, Simba—everything the light touches is our kingdom," Simba's father says.

"Wow," Simba says. And then, "What about that shadowy place?"

"That's beyond our borders—you must never go there, Simba," the king says.

As a young cub, Simba obeys his father and avoids the shadowy place. After the death of his father, Simba sings "Hakuna Matata,"—which translates to no worries—avoiding the darkness while his enemy, Scar, occupies the kingdom and terrorizes its citizens. But as Simba grows up, "he realizes that in order to manage his *inner* map, he must go to those shadow territories, and manage the chaos there," Kabrél explained. "This is a metaphor of what we need to do with our questions. Uncertainty represents these shadow territories that aren't yet incorporated into your kingdom, so you need to be brave, embrace this position of a hero, and start asking questions that will bring light to those dark places. Our enemies—uncertainty, maladaptive emotions, bad habits—become parasites in our minds unless we decide to deal with them."

Kabrél is somewhat of an expert on the subject. He studies the mechanisms in our brain that allow us to uncover intellectual and therapeutic insights. His first published research project started with an observation he made about his own experience in therapy: He often used spatial metaphors when he talked about his internal introspective processes, referring to places where he didn't want to go, dark parts of his psyche (what Carl Jung called our shadow parts[2]), areas unexplored and unmapped, the experience of going in circles.[3]

He wondered, Why did his experience of traveling inward often feel like physical exploration? Why did spatial metaphors come so naturally? Was it the same for other people, and if so, what did that mean?

As he began to learn more, he discovered the neuroscientific concept of mental navigation. It's based on neurological research showing that we use the same brain regions to navigate both mental space—in other words, the relationships between ideas, memories, and thoughts—and physical space.[4] This idea, Kabrél points out, has early roots in philosophy, science, and art; he even cites the idea commonly attributed to Rilke that "the only journey is the one within."[5]

Now that he knew there was a neuroscientific foundation for his query, he still wanted to understand the role that spatial metaphors might play in the self-discovery process. He and a team of researchers compared how much people used spatial language across three contexts—therapeutic conversations, publicly available conversations from podcast recordings, and movie dialogue—looking at approximately thirty-eight million words in total. They also examined 110 psychotherapeutic conversations qualitatively. They found a significant increase in the use of spatial metaphors during therapy sessions as compared to everyday conversations or fictitious conversations, suggesting that the process of mental navigation may be important to the process of therapy and to understanding *why* and *how* therapy helps us to change.

In other words, their research supports a concept that undergirds not only psychotherapy but many spiritual practices: We can change the way we experience our physical and emotional lives through cognitive exploration. These findings also hint at a way we can start to engage differently with our big questions—something we can do to help us feel a little less lost in the haze of uncertainty, unsure of where we are and where we're going.

Uncertain? Start Here

On the following pages, you'll see a questions map. Part art and part science, the map is designed to help you move from feeling stuck in uncertainty to having the clarity to traverse it more purposefully.*

---

* Many thanks to Katrina McHugh, Ariana Wolf, and the entire team at Flight Design Co., who developed the map image for this book.

My hope is that it will help you create your own questions practice, just as you might have one for meditation, yoga, or a creative pursuit. The more I spoke to people who had their own practice, the more I wanted to build my own. But I wasn't sure how or where to begin.

On this map, you'll start by gaining awareness of what your question is, getting curious about whether it's the right question and, if it isn't, what a more helpful question could be. This part of the practice can feel a little bit like being surrounded by trees in a forest. It's a place where you may, at first, lack perspective and feel unable to see the whole picture—a place where it's easy to get, or feel, lost.

Next, you'll consider how to take action—what you want to learn about your question, how you might learn it, whom you'll lean on for support. You're moving from the forest to the mountains, beginning the climb to understand the limits of what you know and, perhaps, to gain new perspective. The action plan you lay out may take some time, even if you've meticulously considered every detail. After all, it can take weeks to acclimatize to high elevations and persevere through mercurial mountain weather. That's why you'll commit to revisiting your question and the map at a particular date.

When you return to the map, you'll have reached an alpine lake. Here, you can take a plunge in the water, eat a peanut butter sandwich, and reflect on what you've learned and where you want to go next. Do you have an answer? Did you identify a new question? Do you want to stick with your original question, or let it go? Will your next steps take you down the mountain, through the forest, or somewhere else entirely? It's up to you.

A map wouldn't be complete without a compass rose, but this one has a different set of directions. Instead of north, south, east, and west, you'll be guided by what I've found are the core elements of a questions practice: curiosity, conversation, community, and commitment. These are the elements we'll learn more about in the next part of the book. No matter where you are on the map, these elements can help you feel like you know where you are and where you're going.

The first part of the practice can be the hardest: identifying your question. Maybe you already have one in mind; if so, that's great. But perhaps you're having trouble figuring out what your question is or whether it's even the right one. You might be sitting with heavy emotions—grief, fear, sadness, anger—and not sure how to face them.

And that raises an important question. How, in the face of our most challenging and painful experiences, can we be curious? This query took me to a strange—and sweaty—place.

START THE JOURNEY

1. What's my question?

Hint: What uncertain situation in your life is making you feel especially curious, anxious, worried, or afraid?

2. Is this the right question?

Hint: Does this question open up possibilities, or narrow them? Is this question supportive or limiting of your growth?

YES — NO

3. What are some other questions I could ask?

Hint: What are some questions that break out of the "yes" or "no" binary, and open up more possibilities for answers?

4. What answers do I already have?

Hint: You may not know all of the answers to your question, but you may already have some knowledge that's helpful. Consider if you've dealt with a similar question in the past.

5. What do I want or need to learn?

6. How and when will I learn it?

7. What could get in the way of my finding the answers I need?

8. How will I overcome those barriers?

9. Who or what else can I draw on for support?

Hint: Who else can you find who may have had similar questions? Or who could help you think through the questions on this map? Consider finding support in other people/communities.

10. Commitment to myself

I will revisit this map on __/__/__ [date] to reflect on what I've learned, and what I may still want to explore.

SIGNATURE:

Reflection / What I learned: ⑱

⑮ **How and when will I learn it?**

⑯ **What could get in the way of my finding the answers I need?**

⑰ **How will I overcome those barriers?**

Hint: It's also okay to take a break. Go to #10 to set a time to return to your question journey.

⑭ **What do I want or need to learn?**

⑬ **What answers do I already have?**

⑫ **What's my *new* question?**

COMMUNITY · CURIOSITY · COMMITMENT · CONVERSATION

CONTINUE THE JOURNEY ➤

⑪ **What have I learned?**

- I have an answer
- I have a new question
- I'm sticking with this question
- I'm letting this question go

  Hint: To make this choice, ask yourself: Is this question moving you toward what you want or away from it?

PART THREE

LIVE THE QUESTIONS →

STARTING A QUESTIONS PRACTICE

7

Reviving Curiosity When Questions Are Painful

THE BIG IDEA

A questions practice is grounded in a mindset of curiosity. That means relating to the uncertainty in your life with more curiosity than fear. By approaching uncertainty with curiosity, you'll start asking questions, allowing you to see your experience more clearly. But how can you get curious about aspects of your uncertainty that might be challenging or painful?

Rilke is in love—again. This time, he's smitten with the painter Elisabeth Dorothée Klossowska, who goes by the name Baladine. Born into an Orthodox Jewish family in Poland, she's spent most of her life in Berlin and Paris, where she married an art historian, had two sons, and first met Rilke. At first, for Rilke, she was merely a "casual acquaintance from prewar Paris days," writes Ralph Freedman, who wrote a biography of Rilke.[1] But when he reconnects with her in Geneva, once again trying to figure out where he will live next, their chemistry is unmistakable. He wanders around the city with Baladine, a dark-haired beauty with a Mediterranean appearance, talking for hours and "touched by a magical attraction."[2]

His love for her is different from what he experienced with the other women he's fallen for in the past, because there is no power asymmetry. Previous lovers appealed to him as either adoring fans of his work or as quasi-mother figures who would shape and protect him. Baladine is neither; she is unfamiliar with his work and is an artist in her own right. She is, in many ways, his equal. He begins to spend more time with her sons, taking an interest in their artistic development that far exceeds the curiosity he's expressed toward his own daughter, Ruth.[3]

But it doesn't take long for him to feel the familiar stirrings, the pull toward isolation. He can't let this relationship get in the way of his work on the Duino Elegies. He'd started the poems, widely regarded as some of his most famous verses, in 1912. He had yet to finish them. When his friends and patrons find him a castle where he can stay for the winter, in solitude, to finally complete the work, he can't resist.[4]

"He had shaken one form of happiness as he opened himself, so he thought, to a more complete fulfillment: his elegies," Freedman writes. "He hoped fervently that his refuge in Castle Berg would help him resolve his painful and unending search."[5]

But Rilke pined for Baladine and was unable to finish the verses; the "experiment in solitude" was unsuccessful.[6] Would he ever find what he was looking for? And what was that, exactly?

It was two days before my thirty-fifth birthday and I was supine on a waterbed in the middle of the afternoon wearing an eye mask. On that scorching late July day, I was sweating profusely. I wouldn't realize until later that after lying on said mattress for about five hours, I smelled more like a farm animal than I ever had in my life.

At that moment, my attention was on something different: I had just taken a powerful dose of MDMA for the purposes of psychedelic therapy,[7] and it didn't seem to be working. MDMA, also known as Ecstasy or Molly, is a substance that affects three major chemicals in the brain: serotonin, norepinephrine, and dopamine.[8] Serotonin can affect your mood and trigger hormones (such as oxytocin) that influence your sense of emotional closeness. Dopamine can make you feel energized, and it's what helps to reinforce new behaviors by creating that positive sense of reward. Norepinephrine increases your heart rate and blood pressure.

Basically, MDMA is supposed to make you feel good. *Really* good. Like the universe is giving you a back massage, a winning lottery ticket, and a plate of warm chocolate chip cookies, all while a bunch of adorable puppies lick your face. But all I felt was numb and anxious. And I wondered, Was there something *very* wrong with me that, even with this powerful substance, I couldn't feel what I thought I was supposed to be feeling: unadulterated love and unfettered happiness? Was this therapy going to "work"? And what would it mean if it did?

A Different Flavor of Curiosity

Let's back up a few months from that sweaty July day. I'd been mapping out the core components of a questions practice and a questions map, from my reporting and research, and knew that *curiosity* was at the start of everything. Questions, after all, flow from curiosity. If we want to live and love the questions of our lives, we must first ask them into conscious existence.

Many people have written entire books defining and redefining curiosity. But for me, the most helpful definition comes from a philosophy professor, Perry Zurn, and a systems neuroscientist, Dani Bassett: "For too long—and still too often—curiosity has been oversimplified," they write, typically "reduced to the simple act of raising a hand or voicing a question, especially from behind a desk or a podium. . . . Scholars generally boil it down to 'information-seeking' behavior or a 'desire to know.' But curiosity is more than a feeling and certainly more than an act. And curiosity is always more than a single move or a single question."[9]

Curiosity works, they write, by "linking ideas, facts, perceptions, sensations and data points together."[10] It is complex, mutating, unpredictable, and transformational. It is, fundamentally, an act of connection, an act of creating relationships between ideas and people. Asking questions then, becoming curious, is not just about wanting to find the answer—it is also about our *need to connect*, with ourselves, with others, with the world.

And this, perhaps, is why our deeper questions are hardly ever satisfied by Google or by fast, easy answers from the Charlatans of Certainty. This is also the reason there is no one-size-fits-all formula for cultivating curiosity—particularly the kind that allows us to live and love our questions, especially the questions that are hard to love, like "How can I live with chronic pain?" or "How do I extricate myself from a challenging relationship?" This kind of curiosity is a special flavor—think of it as chocolate peanut butter brownie fudge, or lemon and elderberry, or caramel with sea salt, or whatever specialty ice cream scoop you enjoy.

It's easy to get curious about queries separate from ourselves ("Why do only some leaves turn colors in the fall?") or even about exciting unknowns in our lives ("What's the surprise my family is planning for my birthday?"). This is vanilla-flavored curiosity—an important flavor, to be sure, but one that's easily accessible in any ice cream store. Just as it's more challenging to find a chocolate peanut butter brownie fudge scoop (believe me, it is), it's also harder to access a sense of curiosity about the parts of our lives that are scary or painful. In the moment, it feels easier to distract ourselves from that pain. Or we may let it hijack us: Our fears and anxieties take the wheel, driving us to places we'd rather not be.

In other words, curiosity can disappear just when we need it most. In fact, some research suggests questions can serve as an antidote to the anxiety that can arise in times of great uncertainty. One study, conducted by researchers at Harvard Medical School and the Pennsylvania State University, analyzed surveys from 6,750 adults that were done approximately once a year for ten years to investigate whether there was a connection between something called "need for cognition" and their symptoms of anxiety and depression.[11] The "need for cognition" (NFC) scale measures how much someone is drawn to the act of thinking—whether they relish deep thought or seek to avoid it at all costs. Though NFC is not the same as curiosity, there are some associations: Generally speaking, the lower your NFC score, the less curious you are likely to be, and the less tolerance you have for uncertainty and ambiguity. After analyzing survey results across ten years, the researchers found that, indeed, higher NFC scores were associated with fewer symptoms of anxiety and depression. One takeaway from the research is that the more you can anchor yourself in questions, letting them fuel your curiosity and deepen your understanding of who you are and what you want, the less anxious you'll be.

When people feel anxious or depressed, they're often wedded to a particular thought or an idea of how they think the future will turn out. This means they're closing themselves off to opportunities, explains Nur Hani Zainal, previously a postdoctoral research associate at Harvard Medical School and a coauthor of the research. "Asking questions opens up different possibilities. If someone can be curious, they are less likely to narrow in on one idea that could be generating that feeling of stuck-ness, or apprehension about the future."[12]

How, I wondered, could I start to approach uncertainty with curiosity? What would it mean to cultivate a unique kind of curiosity—one that's accessible even in the most challenging moments?

Once I started digging, I found all sorts of tools and ideas—ranging from simple to more complex. Many are in this chapter, and I've included more in the appendix. On the simple end are questions you can ask yourself and tricks to bring yourself out of the tornado of anxiety and into the present moment. On the more complex end is guided psychedelic therapy. I'd recommend you pick and choose which tools seem to fit you. And then experiment with them to see what works. You may find that different situations call for different tools and that you're well served by having a robust selection in your toolbox.

Changing My Mind

As a child of the '90s, I grew up believing that all drugs were bad. Full stop. This was the era, after all, of the "Just Say No" to drugs advertising campaign, D.A.R.E. (Drug Abuse Resistance Education), and the

US War on Drugs.[13] Many of these policies were based on outdated or inaccurate ideas about certain substances and disproportionately hurt African American and other communities of color.[14] This led Graham Boyd, founder and director of the American Civil Liberties Union Drug Policy Litigation Project, to call the drug war "the new Jim Crow" in 2001.[15]

But in high school, I was unaware of these issues and conscious only of the ways that I believed drugs like "acid" could corrode my brain. As a leader of an antidrug student group (I was super cool!), I regularly warned students about the dangers of all drugs, including alcohol. If you had talked to eighteen-year-old Elizabeth and told her that, in about seventeen years, she'd be taking psychedelics, she would have laughed in your face.

Fast-forward to me at age thirty-four. By that point, I'd been living in California for seven years and had met person after person who extolled the value of psychedelics for their mental health—including A., who was a huge proponent. Still, I was skeptical. So I tried to learn everything I could—first reading Michael Pollan's *How to Change Your Mind* and then looking at studies from scientists such as Rick Doblin, the founder and president of the Multidisciplinary Association for Psychedelic Studies (MAPS), as well as emerging research from the National Institutes of Health and Johns Hopkins Medicine.[16] What I found intrigued me and began to erode my antidrug zealotry: Researchers were finding in study after study that substances like psilocybin, MDMA, and LSD could heal people suffering from post-traumatic stress disorder, anxiety, and depression more effectively than existing treatments.

For instance: A 2023 meta-analysis found taking psychedelics significantly reduced anxiety and depression for patients with advanced cancer.[17] A randomized clinical trial from 2019 to 2020 showed significant antidepressant effects of psilocybin in adults with a major depressive disorder.[18] And another study in 2018 showed significant anxiety reduction with MDMA-assisted psychotherapy in autistic adults suffering from severe social anxiety.[19]

Critically, the impact of psychedelics on mental health is not positive for all people, all the time. In one 2023 study, researchers found "extended difficulties" resulting from psychedelic use—particularly experiences without a therapeutic guide.[20] Participants reported experiencing feelings of anxiety and fear, social disconnection, and a sense of existential struggle, among other effects. For approximately 33 percent of the 608 participants, these experiences persisted for a year, and for nearly 20 percent more than three years.

Adding to the complexity, not all psychedelics are the same, nor do they all offer the same kinds of benefits (or risks). Though substances like LSD, MDMA, ketamine, and psilocybin are all lumped together under the same umbrella of psychedelics, some have markedly different chemical structures and therefore distinct interactions with your brain.

"The reason they get lumped together is because of the expansive term 'psychedelic' itself," explains Imran Khan, the executive director of the UC Berkeley Center for the Science of Psychedelics.[21] "The word 'psychedelic' comes from the Greek and can be translated as mind-manifesting or 'mind-opening.' In the right set and setting, these drugs, which have very different properties, can lead to similarly mind-opening experiences."

When this book was being written, no direct research on how psychedelics influence our relationship to uncertainty had been conducted. But the experts I spoke to believed other outcomes of psychedelics showed they did have an impact on uncertainty. Natalie Lyla Ginsberg, the global impact officer for MAPS, offered several examples. Many people describe having a spiritual experience on psychedelics, both feeling humbled by the vastness of the world, "having no idea what is beyond," and simultaneously feeling a sense of trust, she said.[22] "Psychedelics can allow people to feel love and safety in the beauty of the great unknown," Ginsberg told me.

Ginsberg, who wrote her college entrance essay on the power of questions, entered the world of psychedelics through her passion for criminal justice work. Growing up in a Jewish family, she devoured

Holocaust fiction, such as the book *When Hitler Stole Pink Rabbit*,[23] as she attempted to answer a series of unanswerable questions: How could something like the Holocaust happen? Why?

"I had this need to understand," she said, learning early on that "asking questions doesn't always lead to answers. There's lots of value in not having an answer."

In college, she knew she wanted to be a public defender but thought it might be valuable to intern at the District Attorney's Office so she could understand the other side. It was during her internship at the Brooklyn DA's Office that she first got curious about drug policy reform.

Ginsberg started asking a lot of questions: Why are so many Black teenagers arrested for marijuana possession and none of my white friends are? Is drug policy being used to criminalize and incarcerate people of color? Why are certain drugs illegal? Are they actually more dangerous than pharmaceuticals? Who wants them to be illegal, and who benefits from them staying that way? Instead of becoming a public defender, she pivoted to social work, figuring that it would allow her to stay more in the gray area of questioning versus what she perceived as the more rigid world of the law. Her next stop: working as a guidance counselor at a school in the Bronx. There, she saw the impact of what she considered racist drug policies and the way that fifth-grade kids were traumatized from police officers "treating them like criminals already." Police would ask kids questions like "What do you have in your bag?" At the same time, she was growing frustrated with traditional therapy as an approach for healing.

"It focused on reducing symptoms rather than understanding the root source," she said. As she continued to explore drug policy, she came across the research on psychedelic therapy, discovering its potentially powerful healing effects, particularly on traumatized individuals. "When you're in a state of trauma and you're in a protective, hypervigilant mode, it's harder to hold multiple truths or ask questions," she said. "Having an answer can feel like safety to someone, but psychedelics can allow more trust in the uncertainty and the unknown." Not only do psychedelics reduce fear of the unknown, but they can help us "actively appreciate that uncertainty creates space for something new. When things are always known and decided it's much more difficult to innovate and move forward."

These were ideas I could understand intellectually. But based on my reading, I thought what psychedelics offered me was a way to know these things *in my body*—to *feel* that they were true. Because as anyone who struggles with their relationship to uncertainty knows, what you may deal with are not just intrusive thoughts but a physiological experience—a racing heart, tight shoulders, a stomach that won't unclench.[24] It's an experience you can't simply think your way out of.

From the aforementioned research, I knew psychedelic therapy came with risks. It could make my anxiety even worse or make me feel even more existentially lost. That said, I also knew that the most common adverse effect was the typically short-term experience of a "bad trip."[25] This could lead to increased anxiety, paranoia and fear stemming from disturbing sensory illusions, disconcerting thoughts or feelings, or strange senses in the body—especially for people who weren't prepared or guided.

After weighing the risks and benefits, I decided the right way forward for me was to find a reputable, experienced guide and to give it a try.

Going Scuba Diving in the Psyche

"What do you want to get out of going deep into your subconscious?" This was the question Chrissy, my new psychedelic therapist, posed to me during our first meeting.[26] The way it would work, she told me, was that we'd do four preparation sessions prior to the journey—the day I would take MDMA. After the journey, we'd have four integration sessions, designed to help me figure out how I would take what I'd learned with me into my life.

I'd been going to regular therapy for years, off and on. As a tool for penetrating my subconscious, though, therapy always felt a bit like beginner's scuba diving with a very limited amount of oxygen. I could swim around, observe some interesting patterns, and then it was time to come up. I needed the steel submersible to guide me down, down, down.

To what, though? What did I want to see down there? Or to understand?

I knew I wanted to explore how psychedelics could impact my relationship to uncertainty. But there was something deeper and much more personal. I wanted to know why I still found it so hard to love myself, to feel genuine compassion and care for who I am, without the contingency of achievement.

"What's holding you back from feeling that way?" she asked me. At first, I couldn't think of a response. *This* felt like the mystery I wanted to solve. Why was it so hard?

Then, some answers popped into my head: It's a habit. The behaviors are ingrained. There are so many little things I do every day that signal I'm not enough, worthy, lovable—like forcing myself to work when I know I'm tired, berating myself for making a mistake, giving in to the pressure to say "yes" to things I know I don't want to do.

There was a complicating factor, though. Some of those behaviors weren't all bad. In fact, I told her, they constituted a strategy that had, at least in my mind, given me a life I loved and appreciated.

This meant that part of what was holding me back was the uncertainty of what could happen if I pursued a different strategy. What would happen if I let those behaviors go?

My fear was that I would lose what I had gained. Things would fall apart. I'd lose control, careening toward insecurity. But I simultaneously began to realize how rare it was for me to explore what could be positive about a shift in strategy. How could it give me an even more fulfilling life? How could it help me become more of the person I wanted to be? And then there was the question that felt like a paradox: How could I find stability and security by letting go of my desire for control?

As I was walking home through the hills, I had a revelation. For so long, I'd imagined that the best analogy for loving and staying committed to a question was our love for others—like my challenging relationship with A. But that wasn't right. My relationship to my questions was a reflection not of my marriage with A., but my relationship *to myself. That* was at the root of being able to exist in uncertainty with more ease and patience. To love the questions, I'd first need to learn to how to show myself compassion. This was an understanding that people such as Mateo and Parker Palmer had hinted at, but now it was finally sinking in.

After returning from therapy, I got a message from my mom on Instagram. It was a link to a short video that asked, "If I asked you to name the things you love, how long would it take you to name yourself?" *Okay, universe!* I thought to myself. *I get the message.*

The Elephant in the Room

"Psychedelics produce these amazing changes in subjective human experience, but we don't really know how they do that," Imran Khan tells me. "Part of the reason we don't know how they work is we also can't fully explain how consciousness works."[27] Researchers *do*, however, have some emerging views on how our brains use sensory input to inform our conscious experience. One of them has to do with the influence psychedelics could exert on our relationship to uncertainty.

To understand this, we need to turn back the clock to understand how scientists *used to think* the conscious brain functioned. Many once assumed that our senses acted like video cameras, passively capturing information from our environment.[28] The brain would then interpret this information, forming an objective representation of the world we experience. But over the past few years, scientists have challenged this model of the brain, suggesting that it's more like a prediction engine.[29] "What it seems to be doing instead is more like a controlled hallucination," Khan explains. "At any given moment, your brain is actively producing a prediction of what it thinks is 'out there.' That prediction is informed by all of your life experience." In other words, when your brain sees a chair, it's not just processing sensory data, it's also channeling all of the other past times you've looked at a similar chair to shape your perception of the furniture in front of you.[30]

Back to how psychedelics might work: One idea, Khan says, is that they shift the balance in your brain away from relying on prior experience (what you're expecting to see) and more toward using sensory experience (the input from your eyes, ears, etc.).[31] This creates more plasticity or malleability in your brain and helps to explain why people can change the way they relate to a mental illness, or substance abuse, or family members during and after a psychedelic experience.[32] "Previously, their brain was relying heavily on top-down priors," Khan says. "An example of a prior might be 'I am a depressed person' or 'I need alcohol to make it through the day.' But under the influence of psychedelics, those influences weaken. The potential for new ways of understanding something are increased, and could potentially create more space for uncertainty and new perspectives in the brain, resulting in different priors."

Research in mice published in 2019 and in 2023 in *Nature* suggests that psychedelics create that space by reopening something called a "critical period" in your brain.[33] A critical period is a neuroscientific concept that refers to times in your life when your brain is more able than at other times to learn and adapt—when it's at maximum plasticity.[34] One example of such a time is when you're learning languages as a kid; another is immediately after a stroke, when there is a need to restore lost cognitive functions.[35]

Khan emphasizes, however, that psychedelics don't always lead to more comfort with uncertainty. The story is more complicated. The drugs may even have the opposite effect, he suggests. We can come to know things in a variety of ways, he explains—by reading about them, by having a firsthand experience, or by feeling like a truth has been revealed. Psychedelics seem to be able to produce the revelatory kind of learning. Revelatory knowledge can create deep certainty about answers to questions—which can be helpful or harmful. A helpful kind of certainty might be realizing that love is the most important thing in your life and recognizing the importance of investing more time in loved ones. A harmful kind of certainty might be the belief that, through the experience of psychedelics, you now have all of the answers to what life is about.

This is a common joke in the field: Despite psychedelics often causing what is known as ego dissolution—a sense of your concept of self falling away and a feeling of being more connected with everything around you—the field boasts many big egos. These are people who believe they "'got the answer' from the experience, and it's on them to convince the rest of the world that their revelation is all we need to solve climate change or polarization" or some other complex world crisis, Khan explains. Psychedelics "do sometimes give people this messianic belief in themselves, which seems at odds with the idea of ego dissolution."

It's a part of the field that "doesn't fit the narrative," he acknowledges. That narrative is the one I initially bought into: that psychedelics raise questions, enhance curiosity, and even help us to become more humble as we experience the world as extraordinary. "Sometimes that is true, but sometimes the opposite is true," he says. "I've seen many occasions where psychedelics lead people to be even more sure of what they always wanted to believe in the first place."

When We're Afraid to Explore Ourselves

Joseph McCowan, a psychedelic therapy practitioner, has spent his life existing outside of a traditional narrative. He has traversed the binaries of identity and pushed against structures—physical and psychological—that sought to divide him.

McCowan, whose father is African American and mother is Italian American, grew up in Los Angeles on the south side of the Santa Monica Freeway. When it was built in the 1960s, the highway razed the homes of many of his African American family members.[36] Nearly every day, McCowan would travel from the south side to the north side, where he went to a predominantly white school. His mother sought better education options for his brother, Justin, who had Down syndrome, and so McCowan got to go to a better school, too. "I grew up looking at a structural divide—the highway—that represents our societal divisions, while also moving through these different worlds," McCowan reflected.[37]

He also saw a different world through his brother. McCowan was the youngest of five, and Justin, who passed away in 2014, was his second-oldest brother. McCowan credits Justin with helping him develop into a psychedelic-assisted therapist, teaching him to see from other perspectives, slow down, be present, and meet people where they are. "I found that when I would do that with him, it would always bring me into a space of wonder and curiosity," he said. "I realized that being slow and deeply engaged was my favorite speed."

McCowan spent college studying psychology and learning about psychedelics, finding that psilocybin and LSD amplified many users' ability to be present, deeply engaged, and connected with their experiences. He discovered that the substances also allowed people to "embrace the questions and the challenges that would arise during that experience," he said. "The not knowing is part of what's really exhilarating in those moments."

But when it came to joining the ranks of the psychedelic therapy movement as a therapist, he hesitated for years. First, the field was predominantly white. Though he knew he could navigate those spaces, he had never felt comfortable in racially homogeneous environments. But even deeper than that, he felt the heaviness of history—the War on Drugs, the pattern of drug researchers experimenting on African American subjects[38]—and wondered how to reconcile his curiosity for psychedelics with these past traumas. Ultimately, he realized he could be part of making the movement more inclusive of different identities, backgrounds, and experiences, and he got licensed as a clinical psychologist in 2016. In particular, he wanted to help people who had experienced racial trauma to heal and to begin to change their relationship to uncertainty and themselves.

Racial trauma, he said, is similar to PTSD; sufferers often experience several traumatic racist encounters over an extended period of time.[39] There's a spectrum of what these encounters might look like: A perpetrator could yell a racial slur, make an assumption about someone's preferences based on a racist stereotype, physically abuse someone, or target someone with racist comments online—just to name a few possible manifestations. Instead of grappling with the pain of their emotions, those experiencing racial trauma can react by avoiding them, pulling away from themselves in the process. It can create "a survival myopia," McCowan said. By dialing back someone's fear response, psychedelics can help the person regain the capacity to connect with and explore themselves. Dialing back the fear response can also allow them to reunite with the world around them, to turn toward it rather than away.

It's a hard process for anyone who has experienced trauma but is distinctly challenging for the African American community. "Black folks in this country don't have a safe relationship with exploration of any kind," he said, explaining how, growing up, he and his family would often be the only Black people camping. It's not that Black people don't like to camp, he said, but historically, in the context of slavery and Jim Crow, they haven't been able to. "Leaving behind the safe zone of community and venturing into the world was not something people felt safe doing," he said. "Entering into a state of vulnerability can be a giant leap to take for people who haven't been able to let their guard down for centuries."

His role as a therapist and as a trainer of other therapists is to help create that bubble of safety so people feel comfortable first exploring themselves. Again and again, he's seen clients form a deeper appreciation for who they are, let go of their struggle to meet societal expectations, and finally start to show self-compassion.

"There's a natural move for many people to embrace complexity more, and there can be a process of bursting our binaries," he said, referring to the experience of breaking out of a simple way of thinking—yes or no, wrong or right. "If you think of a prism, we might be looking through one facet, seeing only white light. But sometimes psychedelic experiences can help us see things in their full color and complexity, including ourselves, and appreciate it."

Substance-Free Self-Compassion

Chrissy and I had three more sessions before the main event—the psychedelic journey. Between meetings, she encouraged me to reflect on a variety of prompts.

How do I know when I feel safe? What would it look like for me to be self-loving outside of using the psychedelic? To me, this was an important one, especially in setting the expectation that the journey was no panacea. It might help me become aware of different truths, barriers, and ideas, but ultimately, I was responsible for integrating those new perspectives into my life.

Part of that integration would mean asking myself questions throughout the day: "What do I need right now?" "How does my body feel?" "What's something I could do to make myself feel safer, or cared for, in this moment?"

"These self-compassion practices might feel a little icky at first," Chrissy warned me, sensing my hesitation. The point, she said, was to experiment with different ways I might show myself care and see which ones would resonate.

As I started to ask myself these questions more regularly during the day—often on walks without my phone—I sensed an answer emerging. What's something I could do to make myself feel safer, or cared for, in this moment? Often, the answer was taking something off my to-do list for the day or releasing myself from some compulsion to push myself harder—at work, in exercising, in doing something for a friend. Feeling safer, I realized, included not just the absence of fear but also the presence of a whole range of other emotions that got eclipsed by a consistent sense of threat: joy, excitement, sadness, disappointment, awe.

Care for myself, then, also meant spending time in emotions that could feel uncomfortable. Sometimes these were positive ones, like the feelings I have when I've achieved a goal or received a compliment. Generally, I'd feel numb in those moments, eager to fast-forward to my next challenge.

"Many people who come from trauma have a 'nourishment barrier,'" Chrissy told me during one session. This happens when the care or "nourishment" you receive at an early age is also mixed with "toxins," according to one therapeutic paper.[40] For instance, if someone has a parent who suffers from mental health issues that make their emotional responses unpredictable—loving on some days and abusive on others—that child could learn to avoid seeking the satisfaction of love and care altogether so they don't experience the negative. As they grow up, they would continue struggling to let themselves feel positive, loving emotions.

As I reflected on other question prompts she shared, I found myself drawn to two in particular: *Who will I be a year from now? And if I could write to myself five years ago, what would I say to prepare that self for what's to come?*

Here seems like the right time to tell you something else about me: I'm obsessed with time travel. Movies, TV shows, books—if it has some kind of time travel theme, I'll find it. And even better if time travel is paired with some twisted, doomed-yet-scintillating romance.

This obsession with time travel extends to my relationship with myself. For the past decade or so, I've had a recurring fantasy: I'm sitting in a coffee shop, reading a book and eating a cheddar chive biscuit (or a savory scone of some sort, depending on the coffee shop—but never mind, I'm getting distracted). An older woman with voluminous, gray curly hair approaches me. "Can I sit here?" she asks, gesturing to the seat across from me. There's something familiar about her.

I say, "Sure, of course." She takes out a book—a favorite of mine, *Unaccustomed Earth*, by Jhumpa Lahiri[41]—and reads quietly for a while. Then she says something like, "Can I tell you something that might surprise you, Elizabeth?"

I look up from my book and biscuit, mid-bite. "How do you know my name?" I ask.

"I know who you are," she says, "Because I am you . . . in thirty years."

I still need to work on this bit of dialogue, which feels a little over-the-top. In my opinion, though, conveying that you are a time traveler is always going to be a moment of high drama.

As I drop the biscuit and ready myself for a speedy getaway, she begins to rattle off things that only I would know about myself: what haunts me when I can't sleep, the name of the doll I lost at a park when I was five, the silly inside jokes I have with A.

I decide to believe her. And then, we talk for hours. The coffee shop closes, so we walk to a restaurant for dinner. Naturally, we order the exact same meal. I ask her endless questions: "How has my marriage turned out?" "Do I have kids—are they healthy?" "Are my parents still around?" "Do I still write?"

She gives me mostly cryptic responses. "I can't tell you that," she says—though she does tell me that in the future products for curly hair have become much better. "Just don't worry so much, okay? Everything will work out."

Unlike most time travel scenarios, this visit from my future self does not, at least in this fantasy, have some larger, cosmic purpose. She is not warning me of some terrible decision I'm about to make, or trying to save a president from assassination, or doing anything else altruistic.

The purpose of her visit is to help me cope with what I don't know today—allowing me to, in a small way, tolerate the uncertainty, because I know that a future me exists.

I find a similar comfort in communing with my past self. As I write to her, I feel a sense of pride and an almost maternal energy. I tell her that the next five years are some of the hardest she'll have to endure, but she'll learn and grow in ways impossible for her to imagine right now.

Why do I find it easier to connect with my past and future self and harder to connect with my present self? I brought this question up in my final therapy session before the big day.

"To me, that's a great sign," Chrissy said, noting that people who are completely lacking in self-compassion often aren't able to connect with any version of themselves. "It suggests that you're really close to being able to feel that way for your present self."

There's also good evidence that this strategy of mental time traveling helps us better navigate uncertainty. The scientific term for it is "temporal distancing."[42] By imagining how a past or future version of ourselves would handle a particular challenge, we can gain more perspective on our big questions today.

An important note: The mental time traveling I did, and the other prompts I explored with my therapist, are tools that *anyone* can use to connect with themselves and cultivate self-curiosity and self-compassion. You don't need psychedelics to ask yourself these questions, and I'm including more of them in the appendix.

The Last Train Stop

The evening before the journey, I felt a little afraid. I'd been anticipating the day for so long and realized that I'd been treating it as a last experience I wanted to have before trying to start a family.

Let me back up for a moment. In the months after A. and I had our reckoning at the top of Mount Rainier, our relationship had slowly transformed. As we picked up our new question—what would have to change for us to stay together?—we began to experiment with answers. We would both need to change the way we communicated with each other. He would need to find ways to challenge me that still felt caring; I would need to surface frustration and concerns before they became festering wounds. And, most of all, we needed to ask each other more questions, to stay curious in painful moments. For instance, when A. seemed to be in a bad mood, in the past I would have taken it personally and closed myself off. His negativity triggered an anxiety that I'd done something wrong or that something was wrong in our relationship. Now I knew to ask, "Are you okay? What's going on?" Usually, I discovered that something was happening in his world completely unrelated to us—something that he wanted to talk about but hadn't been able to yet.

These changes were small, but they accumulated into what felt like a tectonic shift: More than ever before, I trusted the integrity of our relationship. It wasn't perfect, but it felt flexible and strong—so much so that a year after the climb, we started talking about when and whether we wanted to be parents.

And a year after that, we decided we wanted to start trying after my psychedelic therapy.

Now, in my mind, the therapy felt like the last train stop before I plunged headlong into another uncertain journey. As I became aware of how much I'd raised the stakes, I wondered, Could I see this as just another opportunity to understand and accept myself? Or, better yet, could I stop seeing it as instrumental—an experience that had to have a specific purpose? Could it just be whatever it was? This felt like a black belt level of acceptance. I decided to stop thinking so much about it and go to bed.

Unexpected Epiphanies

The next day, when I first entered the therapy room, it was dark, with shag carpets, a large fan whirring, and the waterbed in the corner. On a little table toward the back, Chrissy had placed a candle, incense, dried flowers, and the pills in a small painted bowl.

I swallowed the pills and began the approximately hour-long wait to start feeling the effects. Chrissy left the room so I could journal and stretch. I found that I was thinking lot about my mother and my grandmother. I thought about what it must have been like for my mom to face extreme postpartum depression at a time when it wasn't really talked about,[43] about the pain she felt in leaving me. I was thinking about a conversation I had had the previous day with my grandmother, who cared for me when I was a little girl while my mom was away. My grandmother now has dementia. Before the COVID pandemic, she was sharp and social, regularly seeing friends, playing bridge, and keeping up a lifelong piano practice. Now, she was living in an assisted living center and painfully aware of her own memory loss. I was also thinking about a friend recently diagnosed with an aggressive cancer.

Chrissy returned to the room and encouraged me to lie down on a bed with a satin covering. She tucked me in with a sheet and blanket as I donned an eye mask. I waited, still feeling nothing. And then, suddenly, I had the sense of lifting off. As I felt the effects, I also noticed something else: an awareness of all of the voices inside my head, the parts of me. I remembered the theory of Internal Family Systems—how I was made up of lots of different subpersonalities, many of them created at an early age to try to protect me.[44]

I began naming these parts of me as if they were Santa's reindeer. There was the Timekeeper—the one who voices concerns that there isn't enough time, that someone might be wasting time, that we're running late, that we're too early. *Shouldn't Chrissy be asking more questions?* I wondered. *Shouldn't she be trying to guide me more? What is she waiting for? How long has it been, how long do I have?* And then, the Engineer—the one constantly trying to change my environment, constantly judging things around me. This part of me wondered, *Is it working? Am I high enough? If I was high enough, wouldn't I be less anxious? What the hell is wrong with me? If this doesn't work to "cure" me, what else will?*

Also present was another, more grounded voice that said, *Hey, it's all okay, just try to let go and relax* (maybe this was my inner hippie). I began to notice the anxious thoughts coming in waves and saw the way they interacted. I could observe the conversations between these parts without feeling that they were controlling me.

And then it hit me: These anxious parts of me weren't going away. There was no hack, tip, trick, or substance that would erase them. Even though I knew this intellectually before the journey, I didn't *really* believe it. I still wanted to believe in magic—in fast, easy answers.

But lying on that sweaty waterbed, listening to the cacophonous voices inside of me, I surrendered to a new mindset. All of these sub-personalities were part of me, like family members I needed to learn how to live with. These different parts, I knew, *did* want to help me, even if it didn't always seem that way.

Most of the rest of the session, in my memory, consisted in my rambling to Chrissy, who occasionally responded with a question or "Hmm." The four hours I spent lying on the bed evaporated faster than I imagined, and soon she was telling me that it was almost 5:00 p.m. I was still *extremely* high. A. picked me up, and we walked around our neighborhood as I attempted to recount the experience. I remember him mostly listening and smiling, interjecting only to say, "Can you talk a little slower?"

"I feel like I smell," I told him. "Do I smell?"

"Yes," he said laughing. "You smell very, very bad."

The experience was far from what I'd imagined it would be. What had happened? What would I take from it? Would it change me?

The next day, I noticed that the thoughts and feelings I usually had when I woke up hadn't changed. I still felt the same urge to push myself and grind. What *had* changed was the space I felt between myself and those feelings. Yes, they were part of me, but they weren't all of me. They were something I could be curious about, because I could observe them. I was no longer clinging to them like a lifeboat.

In the *Duino Elegies*, Rilke reflects on how letting go of attachments can create more space. "Fling the nothing you are grasping / out into the spaces we breathe," he writes. "Maybe the birds / will feel in their flight / how the air has expanded."[45]

I found that some of the questions I'd been practicing asking myself before the journey were more habitual now: What do I want and need? How do I feel today? What would make me feel safer?

Finding My Own Space

Over the next few weeks after the journey, I went to four integration sessions with Chrissy. I learned about the tools I could use to create space and care in my mind and body during periods of extreme uncertainty and anxiety—times when the questions became too much to bear.

For instance, we played with techniques from somatic experiencing, a therapeutic approach designed to release trauma from the body. Developed by Peter Levine, PhD, somatic experiencing is based on the idea that when we experience something traumatic, it stays in our body unless we actively release it; much of our suffering comes from lacking awareness of where those emotions are stored and how to let them go.[46] Chrissy taught me how to help myself turn down the dial of my nervous system by realizing that I'm safe.

One practice she shared is called orienting.[47] When you feel yourself becoming anxious or concerned, the idea is to simply pause and look around the space where you are, turning your head to make sure you get the full view. Notice at least one or two things you see that you like. It's a practice that helps your body understand what your mind may already think it knows: *You're safe*. Of course, there may be times when your surrounding environment *isn't* safe, in which case you should leave if you're able to. This practice can be helpful when your environment is objectively a safe one, and yet you're still feeling anxiety and unease.

At one point during a session, Chrissy asked me to close my eyes and describe what I was feeling in my body. In that moment, I told her, I was feeling relaxed.

"How do you know that you're relaxed?" she asked.

"My arms and legs feel heavy," I said. This very basic experience was, it turned out, pretty important. Anxiety tends to thrive in our torsos.[48] When you feel anxious, think about *where* you feel it. Chances are that you feel it in your chest, your stomach, your shoulders, your back, or all of the above. Taking a moment to remember "I also have arms and legs" during periods of discomfort allows us to shift our awareness. We're teaching our body that these feelings and sensations aren't everywhere and that we can move in and out of them.

These practices, in other words, give us freedom and space—space where curiosity can grow. Because anxious and fearful responses to uncertainty feel constricting, closed off, the trick is to figure out ways to create a sense of expansiveness inside of you.* This, to me, was one of the most important lessons of the experience—a lesson that, frankly, I could have learned even without the psychedelic, though the substance certainly helped.

In the weeks after the therapy, I wrapped up my integration sessions with Chrissy and continued to feel a new gentleness in my relationship with myself. Pushing myself hard every day no longer felt right. The voice that told me to do more began to sound grating, like an off-key singer. What I felt was a softening and more ease in checking in with myself; I heard stronger, more compassionate voices chime in amid the anxious static. It was a little like suddenly entering an area with better cell phone service.

When I first read Rilke's words about loving the questions themselves, I imagined my questions as being external to me. They were objects that I might hold, pack with me in a suitcase, place in a locket around my neck, frame next to my bed. And perhaps that is because he writes about them in that way. Love the questions, he writes, "as if they were locked rooms, books written in a foreign tongue." But what I found is that the questions were not things out there in the world. They were deep inside of me, inextricably bound up with who I was and who I was becoming. Perhaps it felt too tidy, too corny, but it was true: To love the questions themselves also meant to love myself.

---

* Researchers have even found that this sense of feeling constricted corresponds with activation in the default mode network regions of the brain, such as the posterior cingulate cortex (which neuroscientists think is involved in memory recollection, among other activities), and that a sense of openness and curiosity correlates with decreased activity in the same area. Judson A. Brewer, Patrick D. Worhunsky, Jeremy R. Gray, et al., "Meditation Experience Is Associated with Differences in Default Mode Network Activity and Connectivity," *Proceedings of the National Academy of Sciences* 108, no. 50 (2011): 20254–9; Marcus E. Raichle, "The Brain's Default Mode Network," *Annual Review of Neuroscience* 38, no. 1 (2015): 433–47.

The Next Phase of a Questions Practice: Conversation

It's possible that the notion of psychedelic therapy still sounds pretty out there to you. Maybe you're not able to get access to it where you live, or you're concerned about the risks. That's okay. It's just one tool among *many* that can help you learn how to approach uncertainty with curiosity (I include additional guidance on this in the appendix of this book).

One of my biggest takeaways from the psychedelic therapy experience was that the journey itself wasn't the only source of relief or insight for me. Everything *around* the journey itself was also important. Sure, I'd changed in part because of the MDMA, but I'd started to feel a shift long before the journey itself. This was the result of the power of practicing my curiosity with myself—asking myself questions and being held accountable by a trusted partner.

For me, this was an important realization. Though questions can exist on their own, as curious solo travelers, they more often come in pairs or groups. They are parts of conversations that we have with ourselves and with others. And these conversations, if we allow them to, can guide us through our uncertainty.

But as anyone who has ever felt tormented by the voices inside their head knows, it's not always easy to have conversations with ourselves, especially when we're already feeling anxious and upset. How do we converse so we see the path forward more clearly, rather than becoming more and more lost?

8

Learning How to Talk to Yourself

THE BIG IDEA

There are lots of therapeutic practices that can help us have better conversations with ourselves. This chapter provides a distillation of those ideas and tools. At the core, most of them use a series of answerable questions to help us feel safer and more secure, so we can be curious about our unanswerable questions.

Rilke knows: He has to finish the poems. It is 1921, and he is forty-five. He started the Duino Elegies—widely regarded as some of his most famous verses—in 1912 but still has not finished them. They are unanswered questions, pulling at his jacket like petulant children.

As he seeks a place to cocoon himself in solitude (once again), he chooses Switzerland—a place he'd once hated but had come to love. Previously, he'd called the Swiss Alps "stupid," no more than "imposing barriers, as senseless as some barred door."[1]

"The fields of snow struck him like a white desert and banal tourist bait," writes the journalist Peter Hulm. "Traveling through Switzerland to Italy, Rilke would draw the blinds of his carriage. He found the scenery of mountains and lakes 'contrived,' with God the stage-manager 'directing the spotlight of sunset onto the mountains.'"[2]

Then Rilke discovers Valais, a southern valley nestled into the Swiss Alps. And he changes his mind.

With burly mountains rising on either side of the rushing Rhone River and the sun streaming overhead, Rilke finds that "the eye is offered a continuous play of the most enchanting transformations, a chess game with hills, as it were."³

He clarifies in another letter that it was not the mountains that compelled him but rather the "peculiar circumstance" that they "create space. Like a Rodin sculpture holds within itself and produces its own spaciousness—this is how, to my eyes, the mountains and hills in these areas of Valais behave. Space emanates from them and flows between them inexhaustibly, so that this valley passage of the Rhone is anything but narrow."⁴ It was "a vista as vast as the world" and "practically biblical."⁵

On June 30, 1921, he and his romantic partner, still the painter Baladine Klossowska, see an advertisement "to rent or to buy" an old thirteenth-century manor on the outskirts of Sierre, a town in Valais. Less than a month later, he moves into the castle, the Château de Muzot, thanks to the support of his patrons.⁶

One of its most important features is its tower—an image Rilke uses again and again his writing. According to a survey of Rilke's work over thirty years, the image of a tower appears more than a hundred times.⁷

To Rilke, the tower is a metaphor for human experience, symbolizing everything from sexuality to spirituality, introspection to retreat, anxiety to ascent, loneliness to healing. The tower is a place where the inner self can have a conversation with the external. Wrapped in the sky and cradled by the mountains, the tower is both vulnerable and impenetrable. Rilke feels a kinship with it, seeking, throughout his life, solitude and connectedness—to be apart and together.⁸

There, in conversation with the internal and the external, Rilke completes his *Duino Elegies* in days—experiencing what he describes as a kind of mental hurricane.⁹

Choose any day of December 2023. They all started the same. I woke up around 5:00 a.m., with a wave of intense nausea. I knew what I had to do, though it went against all of my instincts. In the dark, I weakly grabbed the package of saltines next to my bed and ate three, brushing the crumbs out of the sheets. I sipped some water. Curled in the fetal position for twenty minutes, I considered the Sisyphean tasks before me: brushing my teeth, getting dressed, attempting to eat breakfast. Feeling only slightly less like I was about to puke, I hoisted myself out of bed, often dashing into the bathroom to dry-heave into the toilet.

It wasn't just the nausea and vomiting, which plagued me all day long. It was also the exhaustion—unlike anything I'd ever experienced, nothing like summiting Mount Kilimanjaro or completing a triathlon. One morning, I became winded trying to make the bed and had to sit down to rest twice before finishing. Before all of this, I could barely sit still for thirty minutes without needing to move around. Now, a ten-minute walk felt like a major athletic endeavor. Most days I only had energy to move around our apartment. I recalled sitting in patience researcher Sarah Schnitker's office months earlier, telling her how hard I'd found it to slow down, how much I wanted to be able to do it. *Be careful what you wish for*, I thought ruefully.

And this, I thought ominously, was only the beginning. I had no idea what other symptoms awaited me. What on earth had I signed up for?

Some of you may have guessed what was happening. I was pregnant, to my great surprise. For several months, A. and I had tried to conceive. In the process, I'd wondered and worried about whether I even *could* get pregnant, how long it might take, whether I could get pregnant without any fertility treatments or interventions. When you're one of the last of your friends to have kids, you end up holding a lot of stories of how the whole thing could go. One of my best friends conceived in one week; another friend had been trying for two years, with multiple rounds of IVF and no success. And I heard every other variation in between.

After a few months of trying, and seeking some way, any way, to reduce the inherently uncertain process, I purchased a device designed to measure my hormone levels and give me more insight into my most "fertile days." For the uninitiated, these are the days right before and during ovulation, when the female body releases an egg. (Many women, myself included, have been surprised to discover that you *cannot* get pregnant most days out of the month.) Day after day, I peed on very expensive, seemingly high-tech sticks and inserted them into the egg-shaped gadget. It spat out irregularly low hormone levels, indicating that I had never ovulated. Not good. Meanwhile, I got my period and noticed that my body temperature had been high for about a week, and so had my resting heart rate (thanks, wearable technology!). Given my bleak hormone data, I assumed that these elevated readings were due to stress. I'd been interviewing for a new job and had recently had a challenging visit to see family. But did stress prevent me from ovulating, or something else? I went to the doctor, asking her to review my data. "Is there something wrong with me?" I asked, vibrating with anxiety. "What should I do?"

She stared at me with a mix of fatigue and pity. "If you want," she said, "we can do some basic blood tests, but you'll have to wait until six months to go to a fertility clinic and see if anything is wrong." Then she offered, "I got pregnant when I was thirty-nine. So, there's that." This statement, I gathered, was supposed to make me feel hopeful. But I still felt lost and worried.

The day after I saw the doctor, I signed the new job offer. Two days after that, and a day before A. and I were supposed to move into a new home, I took a pregnancy test. I knew that increased temperature could be a sign of pregnancy,[10] but I was also 99 percent sure that I *wasn't* pregnant. The test was just to rule pregnancy out.

I sat on the toilet in our apartment, piles of boxes and endless moving chores awaiting me outside. Mysteriously, unbelievably, two lines appeared on the pregnancy test. I stared at them for several minutes in disbelief.

"What's going on in there?" A. called to me. "Are you pregnant?"

"Uhhh," I replied. "Well . . ." This did not seem possible. Was this possible?

For a couple of weeks, I was euphoric and terrified, googling every symptom and getting deep in the weeds on safe versus unsafe activities and foods—as if I could control what was happening and what would happen.

And then, around week 6, the debilitating nausea set in. I tried all of the suggested remedies, including vitamins, antinausea bracelets, and ginger tea. Some people, I knew, stayed sick until week 17. Others remained nauseated for their entire pregnancies. At my new job, I was struggling to make it through the day.

"Um hi," I texted a friend with two young kids early one morning. "How did you survive this? I feel nauseous and miserable all of the time and I don't know how I'm going to take several more months of this."

"It's the first whiff of the greatest lesson of motherhood," she replied. "Which is—you will face challenges that you think are insurmountable and then you figure out a way. Because you have to. There's no choice."

There was no shortcut—no fast, easy answer for me to consume.

It struck me that a large part of my suffering was not only from physical symptoms. It was also psychological. Each time I dry-heaved into the toilet at 2:00 a.m., I wondered, *What if I have to deal with this for eight more months? What if this never ends?* I worried that I was going to gain weight too quickly since I had to eat every two hours to keep the nausea at bay (gaining too much weight can lead to complications such as gestational diabetes).[11] I worried about having a miscarriage. I worried about birth defects and complications. I felt crushed under all of the questions and the uncertainty. This, I knew, was what Buddhists call the second arrow. The first arrow is whatever pain or discomfort you experience; the second is what your mind does with that experience.[12]

It was not lost on me that I'd spent the last year studying how to live differently with uncertain situations like this one. What had I learned? How could the wisdom I'd gleaned from hundreds of people help me get through this? The silver lining of this period—which I'll call my Slow Era—was that I found myself with a lot of time to sit and think as I took up a permanent residence on the couch.

The first thing I noticed: Along with the fearful worrier, an admonishing voice in me was chiming in—the perfectionist. Every time I felt depressed, or alone, or miserable, this voice dismissed what I was feeling. "Almost every pregnant woman experiences this," the voice said. "Just grow up. It's not that bad. Stop complaining. You have friends who are going through way worse things. Why can't you just be grateful and suck it up?"

According to this voice, I was not doing my first trimester right; I was never doing *anything* right. *What would it sound like*, I wondered, *to respond to that voice differently, or even pay attention to another voice—a more compassionate one?* I knew from my psychedelic therapy experience that I'd never be able to mute the perfectionist part of me, and that was okay. But how could I talk to myself—my whole self—without getting sucked into an anxiety spiral about all of the things I couldn't control? I had my questions and was starting to grow more curious about my experience. Now I needed to better understand how I could use my questions to help me get clearer or help me move forward, rather than pulling me deeper into the fog of uncertainty.

In this chapter, I've organized what I learned in a series of ideas designed to help anyone who is struggling in a conversation with themselves—strategies that helped me blunt the impact of that second arrow and that may help you, too.

Idea #1: Dropping the Reins

Since she was a little girl, Kate Alcamo has ridden horses—something she admits is out-of-character for someone who relishes control and avoids risks. But her experiences with horses taught her an important lesson that ran counter to her instincts.

"I would always get anxious when I felt the horse was high-energy," as the possibility of getting bucked off seemed higher, said Alcamo, a licensed clinical marriage and family therapist and founder of the Family Therapy Center of Bethesda, in Maryland.[13]

"Your instinct in that moment is to grab hold of the reins and hold their mouths tight, thinking that will help you control the situation, but with horses, the opposite happens. They tend to go faster and pull against you."

Instead, Alcamo said, she learned to "drop the reins." Horses can sense your emotions, so if you're anxious, they'll be anxious, too. Dropping the reins reduces stress for the horse, allows the horse to slow down, and communicates to them that everything is okay. It's the same with the anxiety we might feel over uncertainty in our life, she told me. If we can relax for a moment, and let go of the struggle for control, we will often feel better. "It's often about, How do we not make it worse for ourselves?" In other words, how do we avoid the sting of the second arrow?

This is an approach that's rooted in acceptance and commitment therapy, or ACT, which Alcamo uses in her practice.[14] ACT teaches us that pain combined with *non*acceptance leads to suffering. The more we resist the pain around the uncertainty we're experiencing—dismissing it, pretending it doesn't exist, feeling guilty about it, blowing it out of proportion—the worse we'll feel. At its core, ACT is designed to help us allow and accept what is, instead of struggling against it.

In practice, this first means unwinding a pervasive cultural belief: that feeling uncomfortable emotions means that we're mentally unhealthy or on the wrong track.

Treating mental health in our society has "become such a push to not feel bad," Alcamo said. "Everyone is in search of being happy all the time, and when they're not, there's a significant internal struggle. I tell a lot of my clients their feelings are normal, the whole spectrum of emotion is normal, and it's often our struggle against feeling things like discomfort around uncertainty that causes the problems in our lives."

How do you talk to yourself in the moments when you *are* struggling? Alcamo suggests saying something like "Hey, it's okay that I'm struggling right now and that I don't have this figured out. It's normal and makes sense. It doesn't mean there is a problem or something wrong with me."

Alcamo regularly practices this approach in her own life, where she's also a mom to two girls. But she hasn't always had this capacity and wishes she'd learned about ACT while she was going through IVF with her first pregnancy. It was, she recalled, a time of holding a lot of questions that were largely out of her control and didn't serve her: "When am I going to get pregnant?" "What's wrong with me that I can't get pregnant?" "Will this be the month?" "What do I need to do differently?" Her fertility challenges were unexplained, meaning that the doctors couldn't find anything specifically wrong in the tests they'd done.

Part of the challenge of that time, Alcamo said, was "I didn't know how to just say to myself, *'Right now I'm having a tough time with this. I can only focus on what I can do, and accept that this is where we are.'*" She wishes that she'd been able to distinguish between what she could control—helping herself feel better through lifestyle changes and practicing relaxation—and what she couldn't: her fertility timeline. Later, she realized that she could "take some of this worry I was feeling and turn it into action."

Idea #2: Forget the Myth of "One Right Answer"

Something else that can add to the suffering of uncertainty, particularly when it comes to our big heartwood questions, is the illusion that there's one right answer or one right path, and unless we figure it out, we'll remain lost and in pain.

Dialectical behavioral therapy, or DBT, another evidence-based approach, teaches us that there is no one way to be in or see the world.[15] DBT also says that two seemingly opposing ideas can be true at once and that there could be infinite "right" answers.

Sometimes, it's "the effort to find that one right answer or path that's holding us back," explained Brit Murphy, a clinical psychologist who specializes in DBT.[16] Here, it can be helpful to examine the question you're asking to see whether it's pointing you toward a narrow set of possibilities for answers—for instance, a yes-or-no binary—and consider whether there's a question that could create a greater field of options.

When I was asking whether or not to divorce my husband, I realized at a certain point that I didn't like *either* of the answers: "Yes" meant I would divorce him, which I didn't actually want to do. But "no" didn't seem to acknowledge the issues that had led me to the place of asking the question. I couldn't just answer no and feel like everything was okay. The relationship still felt messy and uncertain, even if we were staying together.

The better question for me, I realized, was more expansive: *How can we continue to be together?* This question offered the possibility of richer answers than yes or no, and it helped us both to think through the changes we would have to make in order to make our partnership work.

Idea #3: Tap into Your Wise Mind

This question—"How can we continue to be together?"—came from my *wise mind*, one of three different states of mind we can experience, according to DBT: the logic mind is a part of you that makes decisions based purely on what's rational—data, logic, facts, rules. The emotion mind is the part of you that makes decisions based on how you feel. And the wise mind is the fusion of the two.[17]

Sounds nice in theory. But when we're anxious or worried about a question, and caught in a storm of stress, it can be hard to balance these two parts.

"If you're emotionally activated, it's hard to feel calm enough to explore big questions and to sit in curiosity," said Marina Kerlow, a marriage and family therapist. "When you're in survival mode, you can't think about hypotheticals or meaningful questions."[18]

During my first trimester, it felt like I was existing in survival mode—always feeling hyperemotional (thanks, hormones!), exhausted, or both. My nervous system was also out of whack because I could no longer regulate it in the ways I used to; exercising, taking long walks, or even meditating sometimes made me feel dizzy. How could I use conversation to help me regulate my nervous system so that I could approach my questions from a wiser and more grounded place?

Part of the answer entailed acknowledging that "conversation" didn't always have to mean words. Instead, it could consist of connecting more fully with my senses, and with the feelings in my body. Kerlow first recommended the "5,4,3,2,1 grounding technique" to help me get out of a place of emotional arousal: the technique asks you to identify five things you can see, four things you can touch, three things you can hear, two things you can smell, and one thing you can taste.[19] Grounding skills, Kerlow explained, can bring your attention back to the here and now. Next, you can start to ask questions to connect with your wise mind, an emotional distress tolerance skill that is part of DBT: *What are the facts I may not be looking at? What do I have evidence for—and what don't I have evidence for? What are my emotions right now? What are they trying to tell me?* A big part of dealing with emotions in therapy, Kerlow said, is acknowledging the feeling that is coming up, naming where it is coming from, and then responding with an action plan to help you get through it.

Though it may seem counterintuitive, the way to alleviate those uncomfortable feelings is often to pay *more* attention to them, not less. "Ninety-eight percent of the time what we get stuck in is what we don't want to feel," says Valeria McCarroll, a psychedelic educator and consultant, trauma therapist, and somatics teacher.[20] In response, we can ask ourselves questions that bring us into deeper contact with those feelings, without needing to explain why they exist or what they mean. The questions are basic: *What am I noticing in my body right now? Where does it feel heavy, light, damp, dry, hot, or cold? Does the feeling have a shape, a color, or a smell?*

Engaging all of our senses to try to describe what's happening in our bodies can help us work with them more creatively, McCarroll says. This is something she does with herself, when she experiences chronic migraines. And it's work that she credits, more than any other method, with helping her heal.

Part of the challenge with any chronic pain or illness is not just the pain but the fear it can elicit: the sense that it's a problem, that it's a symptom of something being wrong; the second arrow. McCarroll has found that if she can just sit with the discomfort and try to locate it, breathe with it, follow it, the pain can move or go away altogether. It is, she says, "an alternative way of working with the energy of suffering." Rather than feeling compelled to tell a story about it (for example, "There's something wrong with me"), she can work directly with how it's manifesting in her body.

If you're dealing with physical agony on top of psychological hardship, McCarroll suggests asking yourself, "What's the story I want to tell about what's happening?" Ten years from now, if you could look back on this time and say, "It was the best thing that ever happened to me," what narrative would allow for that possibility?

McCarroll adds an important caveat to this question. This isn't about negating someone's experience of pain. It's about situating it "within a narrative of self-empowerment and agency" so someone doesn't fall prey to destructive stories of "being victimized by life," and therefore feel powerless in their own healing.

You may feel that asking yourself this question seems absurd: *How could this painful thing I'm experiencing be the best thing that has ever happened to me?* I found this question made for an interesting thought experiment, at the very least. It was more entertaining to think about than "Will I ever feel better?"

Idea #4: Get Some Space from Yourself

Ethan Kross stood in his living room in the middle of the night, gripping a Little League baseball bat, prepared to protect his wife and newborn daughter from a potential intruder. This was a "madman" Kross had never met, who was nowhere in sight. Meanwhile, in the dark, Kross spun around and around the same thoughts "like a horrible carnival ride I couldn't get off. 'It's all my fault . . . I have a healthy, adorable new baby and a wife upstairs who love me. I've put them both at risk. What have I done? How am I going to fix this?'"[21]

The irony: Kross is a University of Michigan psychologist who studies how to control the negative thought spirals that so many of us experience. In other words, he researches how we can tame the "chatter" of our minds and have healthier and more productive conversations with ourselves. Kross defines chatter as "the cyclical negative thoughts and emotions that turn our singular capacity for introspection into a curse rather than a blessing."[22]

How did Kross find himself in the belly of the chatter whale, unable to escape? It started when he'd received a threatening letter in the mail, a response to an interview he'd done recently on *CBS Evening News* about a new neuroscience study showing that the links between physical and emotional pain were more similar than previous research had suggested. The letter, he deduced, came from someone nearby, though he was clueless as to how the research had offended the writer. The police couldn't do much beyond offer him some advice and attempt to quell his fears. They told him he didn't need to worry, that it would blow over, that it was okay to be afraid but he should try to relax.[23]

Unfortunately, this wisdom didn't stick. "Instead, the despairing stream of thoughts running through my head amplified itself in an endless loop," Kross writes in his book *Chatter*.[24] "Should I call the alarm company? Should I get a gun? Should we move? How quickly can I find a new job?" He grew increasingly agitated and unwell, unable to eat and sleep, and finally decided to keep watch with the Little League bat.

Luckily, Kross was able to snap out of it by using the tools his research suggests can help us have better internal conversations—particularly during moments of uncertainty and questioning. As he and I talked, we realized that Rilke was using some of the same psychological distancing techniques when he wrote about the importance of loving the questions to Franz Kappus.[25] Distancing tools—including reframing, imagining you're giving someone else advice, or even saying your name when you talk to yourself—can help us zoom out of our problem and look at the bigger picture. This perspective allows us to step out of the immersive pool of negative emotions that makes it hard to think clearly.

For instance, Rilke was helping Kappus create more psychological distance by encouraging him to step back from his frenzied questioning, reframe his experience, and get some perspective: Kappus couldn't know all of the answers now, and, fundamentally, he might even enjoy the process of getting to them. This exchange didn't benefit only Kappus. Writing advice to Kappus that Rilke *himself* needed to hear was another form of distancing, one that may have helped the great poet grapple with the uncertainty in his own life.

"Uncertainty and lack of control are giant building blocks for chatter," Kross told me. "What Rilke was doing is brilliant in many ways because he took an issue we all struggle with, which is uncertainty, and said, 'Why don't you just learn to love the fact that there aren't clear answers to these things?'" By encouraging a love of questions, "he's giving people reframes to change the way they think about their circumstances, which in turn changes their emotional reaction. You're taking a negative, and converting it into a positive."

Part of what makes reframing so powerful is that it helps us change the stories we tell ourselves, such as the ones therapist Valeria McCarroll referenced. These stories can help us make sense of our questions. They also shape our understanding of who we are. "We use our minds to write the story of our lives, with us as the main character," Kross writes in his book. "Doing so helps us mature, figure out our values and desires and weather change and adversity by keeping us rooted in a continuous identity."[26]

Sometimes we do this well, crafting stories that are supportive and productive, making meaning that allows us to grow and move forward. Other times, our stories can "lead us astray," Kross told me. "That's where understanding how to tell better stories is so vitally important, and where many of the tools I've researched can be helpful."

Idea #5: Put Your Questions in Writing

One of Kross's tools is another that Rilke regularly deployed: expressive writing to make sense of his own life and experience. This writing often appeared in Rilke's voluminous correspondence; he is estimated to have written about eleven thousand letters, seven thousand of which are in print.[27]

James Pennebaker, a psychologist at the University of Texas, has spent decades studying the role that expressive writing can play in helping us to find meaning in our lives and to heal from painful or traumatic experiences.

"Expressive writing is particularly good at helping people deal with uncertainty," he told me, referencing his research.[28] "When I start to write about something that's really bothering me, the reason I'm writing at all is because I'm dealing with something I don't understand. Writing forces me to scratch the boundaries of what I know about. Very often it gives me a new perspective on it. Sometimes the perspective is 'This thing is so big, I don't understand it and I may never understand it. There's no reason for me to ruminate.' It gives me a sense of the boundary [between what I can and can't understand]. That in an odd way can be an answer itself."

When you're stuck in what can feel like a circular or unpleasant internal conversation, expressive writing can help give your thoughts structure, providing clarity once you're able to see the words on a page. In particular, writing about where you've been, or what you already know, can help you figure out where you want to go or what you need to find out. Writing can help you put disparate pieces together into a coherent narrative, allowing you to clear and quiet your mind.

The protocol is simple: In his research, Pennebaker had people write for four nights, fifteen minutes each session.[29] But he's quick to encourage variations. "Experiment and see what works for you," he told me. "The upside is that there's a reasonable chance that it will help you feel better and get on with your life. And the downside is you've spent fifteen minutes doing something that was of no value. If it doesn't work, stop! Try jogging. Go to a bar."

At a time when he was feeling particularly lost, clinical psychologist Maury Joseph took Pennebaker's advice and wrote to himself, in the form of an article, a couple of years after graduate school. During school, he'd been seduced by the idea that he could learn all of the answers and become an expert. Joseph spent his days in the library, reading all of the books he could, fantasizing that when he graduated and became a doctor, he would know all of the things—or at least most of them. "I was subconsciously chasing omniscience," he told me. "But at the time, I was able to disguise that from myself. I was like, 'I'm just reading good books and learning about the field.'"[30]

After school, as he worked with more patients, Joseph pressed up against the boundaries of his own knowledge as he also saw people struggling to tolerate the unknown. That's when he wrote his article "The Discomfort of Not Knowing: How to Live with Unanswered Questions."[31] A lot of people come to therapy hoping their therapist has answers to their questions, Joseph said. And that, he said, is part of a larger problem: a tendency to seek a guru figure or a knowing other—a tendency to fall for the Charlatans of Certainty.

That tendency goes both ways. "[As a therapist], it's comfy to bask in the image of yourself as this knowing person," he said. "But you pretty quickly figure out that if you have a relationship like that with a patient, they aren't going to get better. Because they are still living with the idea that it's possible to be all-knowing, and that they *aren't*. It's a cure based on dependency versus on their own growth, insight, and maturation. They get the anxiety relief of being connected with a guru therapist but have to live in his shadow."

There, again, I recognized the caution that had come from both Rilke and the Quaker activist and writer Parker Palmer: Be careful not to treat anyone—whether they are your partner, friend, or therapist—as your answer or a shortcut to your own development. It's a tempting prospect, and Joseph says that even his awareness of his tendency to pursue answers hasn't made this desire disappear. But the awareness, he said, offers a counterweight against it—along with some other life experiences that pulverized the illusion of certainty.

"A big one for me was the birth of my daughter," he said. He and his wife had developed a birth plan, a concept he calls "such a ridiculous lie. Actual life did what it does and did not go according to plan. The fantasy cracked."

At around 2:00 a.m., after his wife's seventy-two-hour labor and induction, Joseph found himself alone in what was supposed to be the delivery room as his wife was being carted to surgery.

"I looked at the night sky, and it was this weird reddish hue," he recalled. "It was a very profound moment in terms of confirming my smallness, non-omniscience, and non-omnipotence in the world, and I wasn't even the one giving birth. It was one of the times that this idea got through to me the loudest—'Dude, you are just a guy, an animal basically, trying to survive on this planet. You aren't special, you don't get to have a plan.'"

Like Joseph, clinical social worker Kaitlin Nunamann wrote an article to herself as much as to others. Written for *Psychology Today*, it was called "Four Strategies to Soothe Anxious Uncertainty."[32] "I wrote it because I needed to talk to myself while I was going through IVF," she said.[33] "There's so much uncertainty and unpredictability, and a very clear desired outcome." When I spoke to her, Nunamann was expecting a baby in three weeks; the journey had worked out the way she wanted. But the process was not easy. "For a couple of years I lived in uncertainty and this fixation," she recalled. Ultimately, what helped her accept the experience was to stop trying to control it through another diet, exercise, or relaxation tool for the purpose of conceiving. Instead, she would ask herself, *What do I need today to help myself feel more at ease?* For her, it was about "coming back to the present moment and being less outcomes-based with resolving uncertainty."

For both Joseph and Nunamann, the writing was part of the path to get there—to understand their own experience and to offer support to others in an attempt to help themselves.

Idea #6: Stop Asking "Why" and Practice Compassion

Writing stories—or even advice to someone else—can help us make meaning of what's going on in our lives. Sometimes, that meaning allows us to exist with more ease in our uncertainty. Other times, we might be trying to force a sense of meaning that doesn't exist, or that won't help us feel any better. Nunamann, who works primarily with women in their twenties and thirties dealing with generalized anxiety and relationship issues, notes that she sees a lot of people with anxiety asking *why*, in an attempt to uncover hidden meaning. But this approach is not always effective.

"There's this illusion of, if I can figure out why I'm anxious about this, or I'm acting in this way, I won't do it anymore and I'll be okay," she said. The issue is not the self-reflection, but the if-then belief that undergirds it: "If I answer this, then a problem will be solved or something will be fixed." Unfortunately, that's not always how self-reflection works. Sure, sometimes it does help you understand yourself better, but other times it might spark even more questions. In other words, it's a question that does not always lead to answers.

To escape from the "why" trap, Nunamann, like many therapists, recommends somatic techniques to explore what's happening in your body.[34] You may not know *why*, but you *can* know how your breath feels, where you feel tension, that you're clenching your jaw or tensing your shoulders. And you can start to release.

And if you do want to continue the verbal conversation with yourself, she invokes the self-compassion framework popularized by Kristin Neff: First, similarly to Kate Alcamo's advice about dropping the reins, Neff encourages us to identify and accept that *this is a moment of suffering*. Second, we acknowledge that suffering is part of life and, importantly, that we're not alone: Everyone else on the planet experiences it. Finally, we might ask ourselves, *What could I do or say right now to express some kindness to myself?*

Though it might sound squishy, Neff's framework has solid scientific backing.[35] She's found, in study after study, that people who relate to themselves compassionately, particularly in moments of pain, tend to lead healthier, more productive, and happier lives than those who are mired in self-criticism.

Still, that doesn't make her method an easy one to pick up. "For folks who have a trauma history or a negative relationship with themselves, it can be really hard to initiate self-compassion," Nunamann said. "I've worked with people who have said, 'I don't deserve that.'" This reminded me of the nourishment barrier that Chrissy, my psychedelic therapist, had introduced me to. Even if you resist the self-compassion script or it feels fake, she recommends to keep coming back to it. Eventually, it might not feel so strange.

Talk to Someone Else

Part of the way I changed my internal conversation was by acknowledging that I was suffering (thereby practicing self-compassion) and that it was okay to ask for help outside of what I could provide myself. I wanted to have a conversation with someone else—not for answers, but to get another perspective on ways I might improve my own self-talk. Also, I felt lonely. I knew that community and connection were key to getting through moments like this.

I booked a session with my therapist about midway through the first trimester. I told her that many of the people I'd been talking with—the ones who had managed to love the questions of their lives—had done so not by avoiding suffering, but by getting curious about it, sitting with it, and wondering how it might help them grow.

I, too, had been trying to figure out what I could learn from my experience. This felt like a more grounding question—something emerging from my wise mind—than the others that had been ping-ponging in my head, like "What if this never ends?" and "How am I going to survive?" "But," I admitted, "I don't exactly know what this experience is teaching me yet."

As a mother who had gone through a similarly hellish first trimester, my therapist laughed knowingly. "It's also okay to let yourself be," she said. "It's okay if it's just a shitshow right now, and if there's not necessarily a lesson to be learned." This, I supposed, was the part where I let go of the reins, the struggle, the perfectionism; where, perhaps, I stopped asking *why*—stopped asking questions at all—and just allowed myself to be.

It was okay to just sit on the couch and eat the third grilled cheese sandwich of the day. It was okay to walk only thirty steps. It was okay to work less. It was okay to feel sad and afraid, to wish I knew the answers. Even knowing that I had that wish was progress.

9

Holding On and Letting Go

THE BIG IDEA

The final paradox of living the questions is a need to commit to the process of exploring them, while observing which questions may no longer be worthy of our attention. By connecting us with community and creating more structure around our uncertainty, ritual helps us know when to hold on to our questions and when to let them go.

Three years after Rilke completes the *Duino Elegies, he begins to suffer from intermittent physical pain—enough to make him feel he might be dying. The doctors run their tests, and everything looks normal. His symptoms are deemed psychological, but he refuses to be psychoanalyzed, fearful of how an outsider might disturb the sanctity of his mind.*

Ruth, his daughter, offers to visit. He refuses, seemingly because he isn't sure where she would find lodging, but more likely because he doesn't have the energy for a reunion.[1]

For more than a year, he exists in a purgatory between life and death, between certainty and uncertainty. At various points, he lives in a sanatorium, where doctors continue to tell him that he's fine. His biographer Ralph Freedman writes: "He was caught between the reality of pain in his intestines and mouth and the presumption of imagined experience, between the patient's inner certainty that he was breeding a disaster inside his body and the physicians' 'objective' judgment that there was nothing seriously wrong."[2]

For the next year, Rilke continues to suffer, except for a few brief moments when he feels his symptoms lifting. And then, in December 1926, after his fifty-first birthday, the doctors see what they couldn't before: He is dying from leukemia.[3]

In the last two weeks of his life, on December 13, 1926, he writes a letter to his former lover, friend, and confidante Lou Andreas-Salomé, reflecting on the pain that, in his words, encased and supplanted him, day and night.[4] Though he had spent his life trying to find meaning in pain as an artist, this was a form of suffering that seemed to have no lesson buried inside of it.

"Where to find courage?" he asks her.

In this moment of unimaginable pain, he is still curious, asking a question that connects him to her. He expresses both the desire to know and to be known. Where to find courage, indeed?

In his actions are an answer: Courage is in our questions, and it is in our communities.

They were two separate tragedies. Ulrich Baer knew this. But he could not disentangle them. The first tragedy unfolded on a sunny September day in New York City in 2001. Holding his one-year-old son, Baer pointed to a plane in the sky, just as it hit one of the Twin Towers. At that point, Baer, born in Germany, had been living in the US for nearly twenty years. He taught literature and photography at New York University. He was thirty-five years old—exactly halfway between his son and his seventy-year-old father, who was battling cancer.

As fall crept forward and the country grieved the lives of thousands killed in the terrorist attacks of 9/11, Baer's father got worse. One night, Baer had to rush him to the hospital, jarred by seeing his once-vital father helpless, humiliated, and at the mercy of the nurses. On November 22, he died.

For Baer, the timing and magnitude of the events fed off each other, making him doubt that he could care for his son and keep him safe. How could he raise a child in a city where such an attack could happen at any time?

And he had other questions.

"What is the purpose of having a family? Of creating work? I felt like it was all rather meaningless when it could be gone like that," he told me, snapping his fingers.[5] What, he wondered, was the purpose of his life after losing his father and sense of stability? How could he return to being fully alive when it felt as if he'd been violently ripped out of that life? And how to grieve for a paternal relationship that had never lived up to Baer's expectations? He wanted to make sure that he didn't end up having the same kind of relationship with his own son.

Baer was not just any literature professor. He had studied Rilke and knew what the poet would advise him to do in such a moment: to love and commit to his questions. But he bristled at the idea. "I didn't want to love the questions," he recalled; nor did he want to love the state of questioning, the experience of living with uncertainty. "I wanted to reject it, get out of it, find a way to end it. I wanted to go back to feeling normal."

There was, he believed, a path to feeling normal. It consisted of therapy, exercise, and staying busy with work. He pulled himself together for his son, who he knew would respond to his mood. From the outside, Baer looked as though he was doing well, working through his pain. He edited a book of fiction, prose, and poetry called *110 Stories*, an achievement he said provoked his therapist to gush, "I'm so proud of you."[6]

Even after she praised him, he didn't feel any better. He felt more alone than ever. "I did all of these external things to make people feel like I was doing okay, but I wasn't really doing okay. I also didn't have the courage to allow myself to *not* feel okay," he said. The book, the therapy, all of it was a way for him to flee from his questions—and himself.

He fled his questions because he was afraid of what they represented: an emotional black hole that could suck him deep into grief, without any way to escape. "I thought the best thing to do was ward off all emotion, good and bad," he said. "I had this strong sense that if I start crying, I won't know how to stop. I'm also German, so [being unemotional] is innate to me, it's intuitive," he said, laughing.

He didn't want to sit with his questions because he couldn't imagine feeling better without the answers. This notion even felt threatening to Baer: "I thought, what if I can't solve them, and I don't know what to do next?"

As the months slid by, he started to realize that knowing what to do next wasn't the point. He had to find a way to hold on to the questions without feeling that they owed him an answer, a resolution, or a payoff.

"When you love the questions without a promise that they will be resolved, something deeper may open up," he said. Rilke would have approved.

For Baer, the ability to love and commit to his questions emerged in two ways. The first was recognizing that although he couldn't change his relationship with his late father, he could shift the way he showed up for his own son. "I realized I could change my own behavior and grow [as a father] without needing to resolve the other part, which was unresolvable," he reflected. The deeper questions he felt about his father's death, and the insecurity he felt after 9/11, "would stay with me, would always be something I couldn't feel good about. I was able to stay there but also realize there was something in front of me, and I was going to try to do something different. It didn't make the loss better or make me feel less sad, but it made me feel like I could grow in this experience."

Baer wasn't moving on, he was moving *with*—traveling, and living, *with* his questions, *with* the grief. "The shift from engaging with a question as this thing to be tackled and resolved versus this thing that may be unresolved and stay with you—that for me was really hard to do. And even twenty years later, I'm not exactly so well-adjusted that I can go directly into that mode," he laughed.

For the times when he has needed more support to love his questions, when he has, at times, struggled to enter "directly into that mode," he learned a second strategy: integrating ritual-based practices into his life.

Why We Turn to Ritual in Times of Uncertainty

In the most uncertain days of spring 2020, as the COVID-19 pandemic began to ravage the world, comfort emerged from unlikely places. All across Italy, citizens began stepping onto their balconies or rooftops to sing, bang on pots and pans, or play instruments.[7] They hollered out the windows, clapping and applauding the first responders. Opera singer Laura Baldassari trilled from her window.[8] On Friday evenings, after the government broadcast updates on the latest death toll, Italians would perch outside their windows for a boisterous rendition of the national anthem.

The cacophony of gratitude spread to other places: citizens of Paris, Athens, Amsterdam, and Brussels cheered from their windows.[9] Spanish police sang and played instruments for citizens of Mallorca, Spain, in the early days of the lockdown.[10] New Yorkers joined in the singing and banging of pots and pans.[11]

For anyone who studies human culture, this behavior wasn't surprising.

"During times of uncertainty and danger, people often use rituals to reduce their stress and exert control over their environment," wrote University of Texas psychology professor Cristine Legare for the BBC, explaining the appearance of these behaviors at the beginning of the pandemic.[12] And when it comes to being able to live and love our most challenging questions, rituals give us the strength to commit. Sometimes, they also give us a process to let our questions go.

For some of us, the word "ritual" itself may "conjure up monks in dimly lit temples or extremely difficult yoga poses," writes Casper ter Kuile in his book *The Power of Ritual*.[13] The word may also call up associations with other religions. And, indeed, you can find many formal rituals in religious spaces. But ter Kuile's definition, which comes from activist and minister Kathleen McTigue, is meant to help us discover the rituals in our everyday life. These are practices for anyone, regardless of religious or spiritual affiliation. According to McTigue, a ritual must have three components: intention, attention, and repetition.[14] Rituals are predictable—taking place at the same time, in the same way, for the same purpose. They are repeatable. They can connect us with other people. They are a pocket of consistency and meaning in a world that is constantly changing and can feel empty.

And they are everywhere: Think of a family that gathers each year for a holiday meal, a couple who plan a monthly date night, a runner who dons a special set of socks before each race, a grandson who bakes his deceased grandfather's favorite lasagna recipe each year on his birthday, a manager who encourages a weekly team check-in. "The world is full of these rituals!" ter Kuile writes. ". . . We just need to be clear about our intention (what are we inviting into this moment?), bring it to our attention (coming back to being present in this moment), and make space for repetition (coming back to this practice time and again). In this way, rituals make the invisible connections that make life meaningful, visible."[15]

Mike Norton, author of *The Ritual Effect*, agrees.[16] Our life is adorned with rituals, whether we notice them or not. And that is because we live with constant uncertainty. "People turn to rituals in so many domains of life that it's shocking," he told me.[17] "Underlying a lot of them is an anxiety: Does my spouse love me? Am I being a good parent? Will I get over this grief? Am I doing a good job at work?"

Research suggests rituals are also a potent antianxiety tool, acting as a buffer against the stress of uncertainty.[18] We're also more likely to be drawn to ritualized behavior when we're experiencing negative emotions. One study of the Marathi Hindu community in Mauritius, for instance, found that after participants were induced to feel anxious—through preparing a public speech—those who were able to perform a ritual afterward (in this case, praying to deities inside their local Hindu temple) had less perceived and physiological anxiety than those who were told to just relax in a laboratory space.[19]

"Rituals help us feel like we can have more control over outcomes [in our lives]," Norton said. "We use them to calm down and get excited. We can use them for lots of different emotions. But underneath them are questions: 'Can I?' 'Am I able to?' 'Will this be okay?'"

Rituals don't directly answer those questions, but they can provide a moment for us to pay attention to them—which can help guide us toward deeper understanding. "The lack of time is such a huge difficulty for us in living an examined life," ter Kuile told me. "The culture of overwork and exhaustion leaves very little space for people to engage the big questions because they often demand things of us. Ritual is time set aside where we interrupt the productive flurry of every day. They are boundaries in time in which we get to engage the questions that pull at our hearts."[20] For instance, parents with a monthly date night could use the time as an escape from the persistent kid logistics that saturate everyday conversation and instead discuss questions about their identities outside of parenthood, the future of their relationship.

Or rituals can assuage our anxiety for the opposite reason: They're a brief distraction from those questions. Though the word might sound pejorative, distractions aren't always bad—particularly when we're being distracted from an anxiety loop that's not serving us. Rituals can help us pay greater attention to the moment we're in instead of worrying about what's yet to come or what happened in the past.

The Emotion-Ritual Connection

Xunzi, a latter-day follower of the Chinese philosopher Confucius who lived in the third century BCE, extolled rituals for yet another reason. They are, he said, vehicles through which people show love to members of their community. Rituals, Xunzi suggested, allow us to bring our emotions out of our bodies and into the world, to make them concrete, and to use them to connect with the people we care about. Xunzi was particularly interested in funerary rites—how to act, speak, dress, eat, and drink—in order to give proper form to our emotions of grief and sorrow.[21]

The idea that "our emotions are possibilities to be fulfilled through ritual suggests that we cannot regard them as strictly localized in the body," writes the philosopher Curie Virág, who has studied Xunzi's writings. "Instead, we must recognize their interdependence with the environment and with the dynamic situations in which they arise."[22]

This was something Professor Ulrich Baer realized when, in 2004, he started to practice martial arts. First, it was just meant to be an activity for his son—kung fu for kids. But parents were allowed to participate. "It was just tiny four- and five-year-olds and me," he recalled.[23] During one part of the first class, the kids were jumping from a trampoline onto a blue mat. Baer decided he was going to jump, too. He jumped and did a flip in the air, something he'd never done before. It was a spontaneous experience that unlocked a desire in him for more.

His son wasn't interested in kung fu, but Baer joined the temple anyway. He began training with a mostly silent, old-school *shifu*, or martial arts teacher, five or six days a week. And he was terrible. The shifu admonished him, "You have too much language in your head, Uli." The art of kung fu was a form of action meditation, "emptying your head by working out so hard that you can't think about anything else," Baer said. It wasn't just the working out that would come to silence the chatter in his head. It was also the ritual that pervaded the temple: Everything was highly proscribed, from the way you put on your shoes, to how you said hello, to how you began practice.

"For me, living the questions couldn't just be sitting in my room, contemplating the wisdom of poets," he said. "I had to do something with my body. It didn't work otherwise."

Virág believes that we can suffer emotionally because we overemphasize our inner emotions without considering how our behavior, the way we treat other people, and how we spend our time shape how we feel.[24] We sit and stew and analyze rather than act. Rituals allow us to translate our emotions into meaningful action that can, in turn, help to form our feelings and relationships.

And rituals can create what Virág calls "bracketed reality"—spaces where we can, for a short time, take a break from our questions. Virág enters this bracketed reality during conversations with dear friends, which she describes as a ritualized space. "There's a whole protocol to having a good and meaningful conversation," she says. "You find someplace quiet, you look them in the eye, you try to eliminate the distractions, you don't interrupt, you listen, you pay attention. During that time of interaction, anything is possible, and I can banish the longer-term questions. It's so sustaining that [when it's over,] I can go back to the scary reality."*

Rituals offer a sense of control, time to focus on questions, or a space to take a break from them. And there's another way rituals help us move through uncertainty: by deepening our connection to others.

When Rituals Build Community

Sara Josephine Baker was six years old and all dressed up for a special occasion, luxuriating in a white lacy dress with a blue sash, light-blue silk stockings, and light-blue shoes. It was 1879, in Poughkeepsie, New York. She wandered outside her house, admiring her attire and hoping "that someone would come along and see me in all my glory."

Someone did walk by—a girl who, Baker recalled, looked to be about her own age, but who was "thin and peaked and hungry looking, wearing only a ragged old dress the color of ashes."

"It struck me right over the heart," she wrote in her memoir, *Fighting for Life*. "I could not bear the idea that I had so much and she had so little." So Baker stripped off every item of clothing she had on, "underwear and all," and gave it to the little girl. She walked back into her house "completely naked, wondering why I had done it and how to explain my inexplicable conduct."[25]

For the rest of her life, Baker would continually behave in ways that, to many people, seemed inexplicable; her instinct was to question

---

* Virág also told me that she reread *Letters to a Young Poet* every year of graduate school.

what was known and accepted. She was wired not merely to ponder unconventional ideas, but to act on them.

The instinct originated in an unlikely source: her hundred-year-old great-aunt Abby, a Quaker who wore a bonnet with a large gray bow and a handkerchief around her neck. Baker and her siblings delighted in visiting their eccentric aunt, who set her own circadian rhythm—eating breakfast at midnight, lunch at sunrise, and dinner at 11:00 a.m. and going to bed again around noon. Abby would settle herself on the sofa, prop her feet up on a mahogany footstool whose upholstery was embroidered with a parrot, and begin performing stories from her Bible—a "colossal volume which practically smothered her when it was opened across her lap."[26] Baker hung on every word.

And then, Abby set them straight. "Now, children, that is a very silly story," Abby would say after she finished. "I am an old, old lady and I want all of you to remember what I'm saying. It is a silly story and there is not a word of truth in it. Don't ever let anyone tell you that stories like that are true."[27]

Baker recalled, "It would probably be hard to exaggerate the influence that sort of experience may have on a child, learning so early that it is possible to question the unquestionable."[28]

It is possible to question the unquestionable. Later, when Baker became a physician, she was credited with saving the lives of ninety thousand babies because she questioned the medical establishment's long-held perspectives on the proper care of mothers and children. She began pioneering her new ideas about preventive health care in 1908, as the director of the New York City Health Department's new Bureau of Child Hygiene. Before her tenure, the health department's strategy for dealing with sick children was to identify those who were already ill and link them with physicians. Baker suspected they could help more children by intervening earlier. She created a program in which nurses would visit new mothers living in the deadly Lower East Side tenements. Known as the "suicide ward" by some city health inspectors, this was where six thousand European immigrants might be living on a single block and where diseases such as smallpox, typhus, measles, and dysentery thrived in the squalid conditions. As part of Baker's program, the nurse would appear at the door of a mother's home within a day of her delivering a baby, sharing best practices about breastfeeding, stressing the importance of fresh air and regular bathing, and reminding the mother to discourage babies from playing in the gutter.[29]

The program was a resounding success: twelve hundred fewer children died that first summer in the Lower East Side compared with the previous year. Eventually, the city scaled and expanded Baker's program, reducing the infant death rate by 40 percent in just three years.[30]

Baker did whatever it took to save lives. Once, she was called to a tenement to help deliver a baby. She entered the room, teeming with cockroaches and bedbugs, to find the pregnant mother lying on a heap of straw, four frightened children in the corner, and a man on the floor who appeared to be drunk. As she examined the mother, Baker saw a "festering sore" on her back. This, the woman reported, was where her husband "had thrown a kettle of scalding water over her a few days before."

Hearing this accusation, the husband stood up, "crazy with rage, threatening me and her," Baker wrote. She knew she had to get him out of the room so she could deliver the baby. Baker ran into the hall, and the husband trailed her. As he crossed the stairs, she punched him, and he "toppled backward, struck about a third of the way down the rather long stair and slid to the bottom with a hideous crash." Baker returned to the room to deliver the baby and discovered later, with some relief, that she had not inadvertently murdered the man.[31]

Another time, she chased down the infamous cook Typhoid Mary, sitting on her in the ambulance on the way to the hospital so she wouldn't escape. "It was like being in a cage with an angry lion," she wrote.[32]

Baker was also a proud member of a secret club for women who, like her, questioned the unquestionable. Ritual and community were their tools for committing to their most challenging questions.[33]

At first, the women met every other Saturday at the New York restaurant Polly's, filing in around lunchtime.[34] This was the ritual for several years until the mid-1910s, when this secret group, known as the Heterodoxy Club, had to take its meetings underground. Secret Service agents had begun to surveil them at the start of World War I because some members were vocal pacifists, communists, and socialists.[35]

At its peak, the club had about 110 members and included some of the most prominent, accomplished, and notorious women in the world.[36] One of the club members, Crystal Eastman, cofounded the ACLU;[37] others won Pulitzer Prizes or were bestselling authors or pioneering physicians. One, Rose Pastor Stokes, was a cigar maker who, having attended only two years of elementary school, owed her education to the Jewish cigar makers' tradition of paying people to read aloud to them while working.[38] Many members were jailed or beaten for questioning government policies.

Former Unitarian minister Marie Jenney Howe started the group in 1912. It was intended for "unorthodox women . . . women who did things and did them openly," as member Mabel Dodge Luhan wrote in her autobiography.[39] This simple phrase points to something fundamental about the members: Many of them had public profiles, which was still rare for most women, who were often isolated and restricted to domestic spaces. The women of Heterodoxy were journalists, novelists, actresses, activists, physicians, lawyers, and public intellectuals.[40]

Meetings took place entirely off the record, allowing these normally very public women to be vulnerable, voice their ambitions, and have the space to ask questions. What we know about the meetings has been pieced together from the letters and papers of members and their autobiographies.[41]

Heterodoxy Club women didn't just do things in the sense of public achievements. They also did things in their personal lives that most other people weren't doing, such as getting divorced, living with same-sex partners, and having sex outside of marriage. In this way, they were living out big questions that straddled the personal and political. During the club's heyday, divorce was not only rare but also expensive and scandalous, including a trial that would generally land a divorcée's name in the paper. The divorce rate of the US population during the 1910s hovered around 1 percent.[42] In the Heterodoxy Club, it was approximately 30 percent.[43]

According to one member, the club consisted of "59 different varieties of temper, temperament, and viewpoint," and yet, meetings were enjoyable, smooth, and productive because of Howe's diplomatic skills.[44]

Though originally a religious term, "heterodoxy" can describe any ideas—or people—outside the mainstream. Depending on your perspective, heterodoxy could be a form of open-mindedness, an impulse to worship at the altar of truth rather than ideology, to eschew dogma and embrace dissent. Or heterodoxy could be viewed as a threat, serving as a label for dangerous people, ideas, and perspectives.

The women of Heterodoxy saw themselves in the first category; many in society saw them in the second. These women were living lives they weren't "supposed" to be living.[45] And that was the point. In a society just beginning to embrace science and the scientific method, they tested new ideas and experimented in their own lives "to find out how women can best live," wrote club member and psychologist Leta Hollingworth (known for training Carl Rogers, one of the founders of modern psychotherapy).[46] Many saw themselves as amateur social scientists, even if they lacked official credentials. One member, Rheta Childe Dorr, a muckraking journalist, wrote that feminists are better understood as "intellectual biologists and psychologists" rather than as "reformers."[47]

The women of Heterodoxy were bound together by endless curiosity and a desire to question everything—including themselves and one another. Their motto was "The only taboo is on taboo."[48] Howe, the leader, hoped it would be a place where the Victorian straitjacket of politeness and constraint was "outweighed by the sheer delight in honest disagreement and differences which opened the mind to new possibilities, new ways of thinking, living, being," writes Judith Schwarz, author of *Radical Feminists of Heterodoxy*.[49]

By today's standards, the Heterodoxy Club wasn't diverse; it had only one Black member, and the vast majority of its members were upper-middle-class women. But the club was diverse for that era; it had as members Jewish and Irish women, several lesbian couples, and women ranging in age from their twenties to their fifties—not to mention women who had climbed the ranks of predominantly male professions, such as medicine, academia, and journalism.[50]

Like many of the radicals at that time, these women were intimately involved in the labor, birth control, and suffrage movements, and there were sometimes internal clashes. But though many of the women were activists, what made the group unique was that it was meant to encourage exploration, rather than the achievement of specific goals.[51] "It was a place for sharing and community," says Joanna Scutts, author of *Hotbed*, a book about the club.[52]

Often, what women shared and sought community for was their struggle to answer a big question: What does it mean to live a feminist life?

"The idea of being equal to men is easy to grasp," Scutts says. Feminism is clearly compatible, for instance, with knocking down the doors to male centers of public power. "But the question of how it fits into the private, or domestic, sphere was not," she reflects. Increasingly, she says, Heterodoxy members needed to see models of other women attempting to live feminist lives in order to live into this question and forge their own paths. These women needed a community to commit to living their question, especially when it became challenging under the pressure of real life.

Then as now, many of the heterosexual women in the club married men who were publicly outspoken advocates of gender equality. But these beliefs crumbled when their wives became mothers. Take the actress and poet Ida Rauh and Max Eastman, editor of *The Masses*. In the pages of his publication, Eastman wrote and spoke frequently about the importance of gender equality.[53] But once Rauh gave birth, "that threw him completely," Scutts says. "He saw that there was a biological connection between mother and child that he did not have a part in." An old-fashioned belief that still exists today reasserted itself, cracking the foundations of his feminism: Because women knew innately how to care for their children, this meant that they had to be the primary caregivers. Eastman reportedly abdicated responsibility on a biological basis, abandoning his family.

For many men, the biological realities of sex constituted an impenetrable wall on the path to equality: Feminist ideals could go no further. But the women of Heterodoxy weren't convinced, and this is where they relied on one another—often helping members to care for young ones and supporting women who decided to divorce their intransigent partners.

What was radical about the members of Heterodoxy was not just that they were living the question of how to be a feminist in all spheres of life. They also found a strategy to commit to such lifelong, heartwood questions, which are never fully answered: consistently engaging with a caring community. Their ritual of meeting every other week was only a small part of the support they gave one another throughout their lives, which, like ours today, were marked with both micro- and macro-level uncertainty. They lived through World War I, the women's suffrage movement, the Spanish flu pandemic, the Roaring Twenties, and the Great Depression, just to name a few tectonic shifts of that era.

Through the ritual and community of the Heterodoxy Club, members found that they could help one another stay committed to their questions—and therefore, to the lifelong project of learning and knowing themselves.

Releasing the Questions

Like Ulrich Baer's kung fu lessons, neighbors singing and banging pots at their open windows during the pandemic, and the regular meetings of the Heterodoxy Club, rituals born in the midst of uncertainty aren't just valuable for facilitating commitment to questions. Through the space they create for introspection, or as part of their design, rituals can also provide exit routes from questioning, allowing us to release our outlived queries like dead leaves from a tree. Such questions are parts of us that were once alive but may no longer help us grow.

Outside of ritual, you can also develop an instinct for when to hold on to a question—to keep living it—or to let it go. It can be helpful to ask, "Is this question moving me toward what I want or away from it?" If it's not helping you move toward what's important, what's a better question that will help you get to where you want to go?

When Sarah Schnitker, the patience professor from chapter 2, became ill with cyclic vomiting, she had to let go of a question she'd been living subconsciously: "How can I be perfect?" Though she never explicitly asked herself this question, she said it was masked as other ones, such as "How can I pursue excellence?" or "How could this be even better?"[54]

Of course, Schnitker noted, it's crucial to have standards for your work and goals for your life. But for her, constantly trying to level up became toxic. After Schnitker's health problems emerged, she started to experiment with a different question: "What is good enough?" Pregnancy and motherhood offered another opportunity to "keep pushing myself to be okay with good enough, to stop seeking this perfection," she said. What drove her perfectionism in motherhood was what drove it in other areas of life, too: the myth that it would offer her more control. As long as you're the "perfect" mother, nothing bad could happen to your child.

"I remember for a period, when my child was still in her first year, I was like, 'We're not even going to joke about my being a bad mom.' On my worst day I'm a good mom because I love my child, and all of her basic needs are met. Constantly asking 'Am I a good mom?' is not a useful question. The constant evaluation is counterproductive."

Sometimes the decision to let go of a question isn't a conscious one. Parker Palmer, the activist and author from chapter 4, told me that sometimes his big questions "kind of die out on their own. It's not that the questions I was holding when I was in my twenties about my path in life, about vocation, purpose, and meaning, have disappeared at the age of almost eighty-five. But they have changed shape considerably."[55]

For Palmer, letting go of questions is something he determines through feeling as well as thinking. It's a sense of knowing that comes from his body. "The body has its own way of answering questions," he explained, illustrating what he meant with "Body Work," a poem he wrote about such a bodily experience.

Something settled in my body
Or something was settled, I should say
A question, an issue, a quandary unsolved
Dropped from debate into living tissue
And in the body's wisdom was resolved.

The experience of letting go of a question doesn't always feel like the satisfying *thunk* of a resolution settling in your body. It can also feel more like an opening or expansion. Maybe you're holding on to a question about a relationship with someone who is close to you: "Did we make the right decision to break up?" or "How can I have a relationship with someone who hurts me?"

Eventually, the question shifts from something narrow and specific to something broader. It becomes less about the relationship, and more about the emotion at the center of it: how to hold the pain of heartbreak. "You can't think your way out of heartbreak," Palmer said. But we can consider how it might become a creative force in our lives, rather than destructive. We can ask, "Is it embittering me and shattering my heart? Or is it enlarging me, and giving me greater capacity to not only hold my suffering, but other people's as well?" Holding other people's suffering doesn't necessarily mean making yourself miserable on behalf of your friends, but rather being able to empathetically connect with them when they're struggling, and deepening your relationship in the process.

Zen koans are also tools for expanding our questions and creating a fruitful distance between ourselves and our questions so that their own life and meaning become clearer, Joan Sutherland told me.[56] Sutherland is the pioneer of the koan salons that we met in chapter 5.

Take the koan that I'd been carrying with me since I met Sutherland: "Make the mountains dance." Within it was a central question—How do you make mountains dance?—designed not just for abstract thinking, but to be held up alongside the questions I'd been holding in my life. For me, these questions were: How could I move more flexibly, even joyfully, with the heaviness I felt in my life about seemingly unsurmountable uncertainty? Instead of seeing that uncertainty as something separate from me, like mountains in the distance, how could I come to see it as part of me, as something I was in relationship with, and therefore had power in shaping? The koan's instruction felt directed to me: *I* was supposed to make the mountains dance. And the only way I could do that was by rethinking *where* the mountains were and *what* the mountains were.

The koan invites you out of the realm of the known and into one of mystery. If we think about each question as a room, Sutherland said, we could imagine the koan as opening doors or windows in that room. When you hold a koan alongside your question, or even replace your question with a koan, "it's a way of making the question about more than you, about letting the question itself speak. A good question will always make the room bigger, or open a door in the room to let something else in, or blow down the walls of the room entirely."

Another way to think about whether to hold on to or let go of our question, then, is to ask, Is your question making you feel small, with an ever-narrowing sense of possibility? Or does it make you feel expansive, offering many ways to move forward?

Just as we have a tendency to cling to fast, easy answers in the face of uncertainty, we can also grip certain questions too tightly, limiting our ability to grow and giving the question all of the power. In a strange way, by holding on to our questions for too long, we're also pursuing the illusion of certainty: The longer we hold on to the question, the greater the chance that it will, eventually be answered. In her book *Quit*, social scientist Annie Duke writes about this conundrum when it comes to deciding whether to persevere or walk away from anything in life: "The desire for certainty is the siren song calling us to persevere, because perseverance is the only path to knowing for sure how things will turn out if you stay the course," she explains. "The problem, of course, is that sometimes, the siren song lures you toward a rocky shoal that breaks your ship apart." Though the idea of knowing for sure can be enticing, staying for too long with a question—or a person, an investment, or a job—"is when you lose ground," Duke writes.[57]

Again, letting go is easier said than done. We humans are averse to losing anything, including the answer to a question that may have been useful or comforting for a time but no longer is. Maybe it feels as if letting go is also losing all of the effort you've put into finding an answer up to this point. Loss aversion blinds us to what we can gain from release.

Building Your Own Questions Ritual

Schnitker finds it easier to understand when to hold or release questions by locating where she is in her questioning journey.[58] To do so, she occasionally consults a model of spiritual dwelling and seeking developed by Steven J. Sandage, professor of psychology and religion at Boston University, and F. LeRon Shults, a professor at the University of Agder.[59] I've found the model useful for anyone—even someone who doesn't identify as spiritual or religious—who is curious to understand the rhythm of questioning that we all encounter throughout our lives. The model is an important reminder that we need the constant interplay of doubt and certainty to "make living the questions sustainable," Schnitker said. "If every day [you're grappling with] an existential question, that will get hard to sustain without serious mental health ramifications. It's very taxing on your system."

Instead, thinking about how we can cycle back and forth between "dwelling" in safety and seeking in uncertainty "allows you to live the questions well during the period when you're seeking," Schnitker said. In other words, if you're feeling exhausted from a time of intense seeking, that might mean you need a break. It doesn't have to be days or weeks. It could be an hour or two of relaxing in nature or reading a book that allows you to pause the question. Many religious practices, Schnitker points out, build the rhythm of seeking and dwelling into their calendars of rituals and celebrations, marking seasons meant for celebration and others for questioning or penance.

Take, for instance, the Jewish holiday of Passover: Through its rituals, the holiday promotes asking questions and challenges dogmatism while it celebrates the physical liberation of the Jewish people from Egypt. It's followed seven weeks later by Shavuot, a celebration of the harvest of spring and the time when the Jewish people received the Torah on Mount Sinai. Or, in the Christian calendar, consider the period of Lent, which is meant as a reminder of the forty days Jesus Christ spent fasting in the desert before starting his life in the ministry. For many Christians, Lent is a period of self-reflection and examination, followed by the celebration of Christ's resurrection on Easter. In the Hindu tradition, there isn't a specific holiday or ritual celebration that reflects this same rhythm, according to Dr. Vineet Chander, an assistant dean in the Office of Religious Life at Princeton University, who serves as the Hindu chaplain on that campus. Even so, Chander said, "Questioning and penance are vital parts of Hindu practice and tradition. In fact, many Hindu scriptures explicitly take the form of dialogues marked by question, challenge, response, and further questions."[60] He named the Bhagavad Gita, an ancient Hindu scripture, as one of the most famous examples, as well as the Vedanta Sutra, the codification of Hinduism's popular school of metaphysics. The Vedanta Sutra begins with the aphorism *athāto brahma jijñāsā*, which has been translated as "now is the time to inquire into the Divine . . ."

Even Rilke acknowledged the importance of experiencing moments in which we feel grounded and settled, Ulrich Baer, the Rilke translator and literature professor, told me.[61] For Rilke, natural beauty, mystical experiences, and romantic and erotic love could create "moments of timelessness when we aren't calculating how long we will be there, but we are just present and aware," Baer said. For him, those moments had to exist on their own; they were not cultivated for the purpose of energy or fuel for the next thing.

Starting Your Questions Map

Developing a questions practice, starting with the questions map, is one tool to help you commit and get curious. I asked ritual expert Mike Norton about what would help make the questions map into a ritual—something sticky and supportive through uncertainty.[62] He named a few key characteristics: Whatever questions you use in your reflection, make sure they're the same questions each time, in the same order. It's important to see the underlying logic and flow of the questions you're asking if you want to believe that it might have a benefit. You want to try to give yourself a sense of progress through that preordained set of steps—that you know clearly where you are and how much further you have to go. It can be important to feel ownership over the ritual itself—so, perhaps add a question or two that resonate with you or remove the parts that you don't like. Other than that, he said, there's really no formula.

I would add: Consider filling in the map with a friend or a group of friends. If you can, try to let go of expectations of what outcomes it might produce. Sure, we may grow, change, and become better as a result of a commitment to our questions. But we don't get to determine the timeline of that growth and change, as much as we'd like to. Though we can predict the steps of our ritual, we can't be certain what it will lead to. This is part of the challenge and the joy. Perhaps you will, in Rilke's words, "gradually, without noticing it, live along some distant day into the answer."[63] Or maybe you'll discover even more satisfying questions along the way, questions that give you the courage to move through and with the uncertainty you face.

EPILOGUE

Where to Find Courage?

One January night, on what felt like day 700 of pregnancy, I lay down to sleep. Minutes later, I sprang up, bolted to the bathroom, and threw up my entire dinner.

As he always did, A. had followed me into the bathroom, rubbing my back and wiping down the toilet.

"Do you need any tea?" he asked me. I was sitting on the floor, trying to breathe and stop dry-heaving. I shook my head. A few moments later, I stood up shakily and returned with him to bed.

"Just wake me up if you need me, okay?" he whispered as we crawled under the covers.

As I tried to fall asleep, I thought back to how many nights I'd nearly made myself sick questioning whether he was the right partner for me, whether we should divorce, whether he cared about me at all. In this case, I'd lived my way into an answer. We both cared deeply about each other but hadn't always interpreted the others' actions as care. I didn't want a divorce but rather to learn *how we could both change* to become stronger as individuals and as a couple. Our relationship wasn't perfect—whose is? But after months of difficult conversations, the foundation felt strong.

Now new questions were sprouting. How would A. and I parent together? How would I change from this pregnancy? How could I continue to nurture my relationship with myself while caring for my baby? There were hundreds of these queries—some answerable, some not—that were swimming toward me, like fierce triathletes.

Someday, I imagined the answers to those questions would be stored away in my core, my heartwood.

But that was not today. Today I had to face them, walk with them, just as I had done with countless questions before them. Today, on the precipice of both an ending and a beginning, I had to grapple once again with not knowing what would come next.

I recalled Rilke's last letter to Lou Andreas-Salomé, and the final question he asked her: *Where to find courage?*[1]

Epilogues and final chapters carry a lot of pressure for writers. They contain an unspoken social contract with readers: It's time to wrap it up. Serve up the final course, the satisfying answer, resolution, or conclusion.

We want this at the end of a book and at the end of any experience—a sense of meaning, what it was all for. Rilke wanted this, too, toward the end of his life, when he hoped that inspiration would strike him at the Château Muzot in Sierre and he would *finally* finish the poems he had started a decade earlier.[2]

One day in late June, more than one hundred years after Rilke's Swiss journey, I rode the train from Geneva to Sierre. I was eager to inhabit the place that had proved transformational for Rilke, wondering whether I, too, might experience the kind of creative hurricane and resolution he had as he finished his *Duino Elegies* over several days in February 1922.[3] What, I wondered, could Rilke, and this place, teach me about transitional moments and how to find courage inside of them?

As we sped past castles, horses, cows, terraced vineyards, lakes, and cottages carved into the side of the hills, I saw a landscape lush with paradox. The mountains pressing up against the train were both formidable and soft, imposing and graceful. In summertime, the environment managed to be both serene and searing, gentle breezes offering brief relief under a scorching sun. I understood what Rilke meant when he wrote, of the climate, "there is almost a tenderness in the midst of this severity."[4]

The paths etched into the sides of the Alps looked like elaborate labyrinths. I could see snow hiding in the crevices of rocks, the remnants of winter, which felt especially distant in the sweltering valley. It was a landscape itself in transition.

I'll admit it: On the train, I was tempted to send some emails—to look not at the "practically biblical" landscape unfurling outside my window, but at my laptop screen.[5] This was the fourth leg of a long trip across Europe, and I felt behind on my writing and research. As I felt the mechanical pull of my productivity anxiety, a voice inside whispered, *Be here now. Look at what's around you. What would Rilke do?*

Well, I reasoned, Rilke might actually opt for working. Nevertheless, I shut my laptop and looked outside. I let myself witness the beauty and marinate in the discomfort of feeling hopelessly behind.

When I arrived in Sierre, my first stop was the Château de Muzot, Rilke's former residence.[6] To reach the manor, nestled in the hills about a mile up from town, I rode an e-bike lent by the hotel where I was staying. Though I was haunted by memories of the Bali motor scooters, I figured I could handle that.

Because the Château is privately owned and closed to the public, I stayed outside the gates. I snapped some photos and chatted with one of the French-speaking neighbors—in Spanish, the only language we both shared. He seemed excited that someone from the US had traveled there. It seemed as though he would have been content to spend another hour chitchatting.

This unhurried quality of life trailed me into the evening. At a restaurant, I saw guests enter and, upon seeing friends, spend the next half hour communing with them before sitting down at their own table to eat a meal. When I returned to my hotel, a refurbished castle and winery, I perched outside to watch the sunset. It was 9:00 p.m., and the sun wouldn't set for another half hour. The mountains in the distance looked hazy, as if they were a painting coated in a layer of dust. In the breeze, with the buttery light, leaves became loose and fluttery, making the trees look like a natural disco ball.

The next day, I took the e-bike out once again, this time heading to the Rilke Museum, the only one of its kind in the world. There was one other man who entered at the same time, right after it opened. Otherwise the museum was empty.

None of the exhibits had English labels, which I'd suspected might be the case. Luckily, a staff member gave me a book with English translations so I could follow along. I entered the first room and was flipping through the book when the man approached me.

"Excuse me," he said in accented English, looking at the book with amusement. "Maybe I can help you. I know a lot about Rilke. I can help to translate?"

I learned that he was not just a Rilke fan but the current resident of the Château de Muzot. The widow of its former owner, now in her nineties, allowed poets to stay in the Château for a week at a time. Around six years ago, this man sent her his poetry, and she invited him to stay, too. He had been traveling to the Château regularly over the past five years from Munich, where he was a social worker. The Château, he told me, is his paradise. He had been reading Rilke since he was eighteen and was most enamored with Rilke's eyes—his powers of observation and truly *seeing*.

"Would you like to see the Château?" he asked, his own eyes wide and hopeful.

Bien sûr!

Stepping inside the Château was like entering a time machine. Other than the addition of running water and electricity, which Rilke's manor lacked, most of the rooms appeared to be untouched from Rilke's time, with one exception: In the kitchen, a refrigerator was decorated with drawings by the current owner's grandchildren. There was the grand wooden table where Rilke ate his meals, built in 1688, according to the date carved into the side; the humble standing desk where Rilke wrote his famous poems; and a second desk where he corresponded with hundreds of friends and fans. By the window sat a worn red couch and some of his artworks, religious images from Spain, according to my tour guide, hung on the walls. I paged through the massive, ancient French dictionary he had consulted and gazed at the view from his window. Standing on the balcony where he spent hours observing the garden, I looked out at the terraced vineyards, and the robust ginkgo tree that, when he was there, may have been just a sapling.

I could see how a place like this—enveloped in mountains and sky, embroidered with history—could evoke the kind of creative storm Rilke had experienced nearly one hundred years earlier. The dramatic environment beckoned me outside of my head and into the world, fastening my experience to bigger and universal truths. I could see, too, how this secluded place could be lonely, motivating Rilke to obsessively write letters to his many eager correspondents.

"Where to find courage?" Rilke had asked Lou Andreas-Salomé. How could he face the uncertainty of his last days, the inescapable pain in death and endings?

Rilke had found courage before—found it in the patient way he waited for the *Duino Elegies*, poems that would define his career. The poet Adam Zagajewski describes "the most mesmerizing part" of his biography as his "iron-willed waiting for the *Duino Elegies* to arrive, to visit his poetic mind."[7] When, at last, the poems rapped at his door, they "gave a glorious meaning to his entire pilgrimage, to his waiting, to his procrastinating, to his moving from one villa to another, to his patience. They give Rilke's life the shape of a work of art, made him into a twentieth-century emblem of poetry."[8]

But now, seemingly, what he'd found in living that question was lost again.

Or was it?

In writing this last letter to Andreas-Salomé, his confidante of decades, he may have been subconsciously delivering an answer to his query: The courage to endure any life change emerges from our relationships with others, and our ability to see that we are part of a much larger whole. Finding and nourishing these connections can help us see that we do not need to be alone with our questions. Navigating uncertainty can feel isolating, but it need not be. Instead, it can inspire us to engage in a lifelong conversation, continuously discovering new participants and perspectives.

Too often this reality is hard to see, a myopia that Rilke wrestled with for much of his life. In his "Notes on the Melody of Things," which he wrote at age twenty-three, Rilke wrote about the power of deep pain—such as the loss of a loved one—to bring people closer together. These moments of finality remind people that, as humans, they are all part of an "expanse of melody . . . woven of a thousand voices."[9] Without this deep pain "to make people equally silent," some never take the time to listen and hear that powerful melody, he writes. "They are like trees which have forgotten their roots and now think that their strength and life force is the rustling of the branches."

These people, he writes, "have lost the purpose of existence. They strike the keyboard of days and play the same monotonous diminishing note over and over."[10] Not only do these people forget the deeper strength they can find in relationship to others, they also delude themselves into thinking that all they want are simple, predictable experiences—to hear one note rather than a soaring symphony or a velvety jazz quartet.

In the midst of our most uncertain moments, when we crave sameness, simplicity, and knowing, it takes courage to stop playing the single note, to look around and listen.

For me, conversations like the one Rilke had with Andreas-Salomé had become a reliable way to stop my own mechanical keyboard playing. These conversations, rooted in meaningful questions, helped me discover new reservoirs of courage. The best ones pushed me to reconsider my questions or seeded new ones. Take, for instance, the days-long conversation A. and I had when I shifted my question from *Should we be together?* to *How can we be together?*; or when the astrologer Jessica Lanyadoo helped me see how I was asking the wrong questions of the wrong source. Consider my conversations with Chrissy, the psychedelic guide, who asked me questions that changed the way I related to myself. Or even the "motivational interviewing" session in Sarah Schnitker's office, when she helped me see, through a powerful new question, what it would take for me to become a more patient person.

Others have found courage through these kinds of exchanges, too.

Recall Miguel's conversations with other undocumented immigrants on the parole trip and how they illuminated a different way of living with uncertainty. Or the way Mateo healed in large part through discussion with others in his peer support group. Or how Megan Rundel started to feel more grounded in who she was and what she wanted through the koan salons.

We had all found courage in this ancient art of talk, entwining questions, responses, and silences into something greater and even more meaningful than certainty. These conversations did not always transmit answers, but rather *knowing and familiarity*—a clearer understanding of who we are, what we want, and how we are all more similar in that becoming and wanting than we might imagine.

Months after the European trip, I learned something else about courage from another conversation: It's required of us not just to endure the painful days of uncertainty but the joyful ones, too.

Pregnant and slogging through a gray and drizzly neighborhood walk, I called my mother to distract me from constant queasiness and the gloomy weather.

"How have you been feeling?" she asked me. I updated her on the nausea and nighttime puking, but then she pushed a little more. She meant how I'd been feeling about the baby—about becoming a mom. After her own postpartum experience with me, she knew how complicated the answer to that question could be.

"Honestly, I've been feeling afraid of feeling anything," I told her, explaining that until I was further along, I worried about becoming too attached to the being growing inside me in case something happened. As I said it, I realized how ludicrous it sounded. Too attached? The baby was literally *inside me*. Too late for that.

"I understand that you're feeling that way," she said empathetically. And then, she asked me something that startled me. "But . . . what if you allowed yourself to feel joy, too?"

At first, I was a little defensive. *She doesn't get it*, I thought, as I told her that I'd think about it. I realized later, *She does get it. She gets it more than I do.* I chuckled a little bit, humbled once again by all that I could still learn on this quest. By keeping my feelings at bay, I was continuing to cling to the illusion of control and certainty. If I could just numb myself to emotions now, said the voice in my head, I wouldn't feel any pain later.

But I couldn't control what happened later. I couldn't protect myself from what I might feel or what might happen in the future. I was living a set of questions about pregnancy and motherhood. Some I would answer, eventually. In the meantime, there was something I could do today.

I could stay open to new questions and allow my uncertainty to change shapes.

What if I allowed myself to feel joy, too?

Courage, I'd found in the course of countless conversations, was also in the questions themselves. I'd found courage in my questions practice, using queries to find my way not to certainty, but to greater clarity. And I'd found it in sharing questions with others, discovering that they, too, were caught in webs of doubt. I'd found it through my relationships, in quiet moments in nature, on walks, and in therapy. If questions were indeed like locked rooms, it took courage not only to go searching for the key, but to open the doors and wander inside.

I'd found courage before. Sometimes, like a key, I misplaced it. But I had to remember: It was never lost forever. It was always hiding in some familiar yet surprising place—ready to help me unlock the door and take those first few steps into the unknown. It was just as Rilke wrote when he was, himself, at the start of his career and life:

"I can imagine no knowledge holier than this: that you must become a beginner. Someone who writes the first word after a centuries-long dash."[11]

ACKNOWLEDGMENTS

The very best part of writing this book was not the writing part. It was the conversations it afforded. I had an excuse to talk to some of the most interesting, kind, and insightful people I've ever had the fortune of knowing.

The people whose stories you read in these pages genuinely inspired and buoyed me during tumultuous times in my own life. They have enriched this book, and my life, immeasurably. I thank each of them for their generosity of time and spirit, and for sharing some of the most intimate details of their lives so that we all might know ourselves a little better.

Arriving at those conversations took work. Not just from me, but a whole team of people who believed in me and this project. Margo Beth Fleming, my agent, saw something in my nebulous desire to "write a book about questions" from the very beginning, teaching me how to transform an idea into a *book* idea and keeping me laughing along the way. She has helped me live the question of what this book could be, while consistently challenging me to sharpen my approach. Margo, I'm so lucky to have you in my corner.

Margo introduced me to Heather Kreidler, fact-checker extraordinaire, whose meticulous work and enthusiasm for the project has helped me sleep more soundly at night.

And I would never have met Margo if it weren't for the team at *Behavioral Scientist* magazine, especially Evan Nesterak, Mitra Salasel, Dave Nussbaum, and Cameron French. Evan, thank you for responding to my cold email all of those years ago and letting me be a part of the beautiful publication you have grown and nurtured. Mitra, Dave and Cameron—you have all been a huge part of encouraging me on this journey, offering your thoughts and ideas at various junctures and introducing me to your contacts. I am so grateful.

Daniella Wexler was my initial editor at HarperOne; I'll always remember the first conversation I had with her, where she floored me with her penetrating questions. When she left to become a therapist (go, Daniella!), Rakesh Satyal took over. What an encouraging, incisive, and talented editor he has been, managing to always see where the manuscript could be improved while guiding me with such care through the process.

Speaking of editors, I was lucky to have three exceptional others looking over my shoulder at various points along the way. My father, a longtime journalist, provided invaluable edits and commentary on drafts of the book. My mother, an author and journalist, was another crucial early reader and editor. I have them to thank for treating me seriously as a writer from the time I was eight or nine and already stealing away to write stories.

And my husband, the third editor. As a talented fiction writer himself, he played a critical role in excavating the book's narrative and coaching me through the highs and lows of the creative process. But even more than that, this book would not have been possible to write without his continuous love, support, and questions. I can't say it any better than the great JT: "You are, you are, the love of my life."

Finally, to my son, still so little as I write this. Thank you for waiting to be born until I submitted this manuscript to production. What a considerate thing to do! I can't wait until you ask your first question.

APPENDIX

Here are a few more tools and ideas from leading researchers that can help you develop a more curious relationship to the anxiety that can come from uncertainty.

Break the Habit Loop

In his book *Unwinding Anxiety*, Judson Brewer, MD, PhD, outlines other ways of altering the reaction we have to our anxious thoughts and feelings.[1] His work, based on decades of research, shows that feeling anxiety is a *habit*, which means we can't just think our way out of it. We keep doing habits that are inherently rewarding, so part of breaking the anxiety habit is recognizing that (1) we are anxious and (2) anxiety is *not* actually helping us. Another part of breaking the habit is being able to step outside the habit loop for a moment—to separate ourselves from our anxiety so that we can observe it.

In this space, we can introduce some curiosity: *How is this feeling serving me right now?*

"Curiosity is a key attitude that, when paired with awareness, helps you change habits—a connection backed up by research done in my lab and by others," Brewer writes. "The attitude of curiosity is about bringing a playful attitude to the thoughts and emotions that arise for you."[2] Curiosity itself is a reward because it feels better than anxiety, he suggests.

Practice a New Mantra

Brewer suggests saying "Hmmm . . ." when you feel anxious thoughts or feelings come up in your body. It's a tool to bring yourself into the present moment, and to create the space for curiosity. From that "hmmm . . ." you can start to ask yourself some questions: *Where do I feel this in my body? What does it feel like? Have I felt this before?* That simple awareness, he's found in his research, can help mitigate the intensity of anxiety symptoms—"The restless, driven quality of DO something!" he writes.[3]

Watch Out for Distraction

There's a distinction between curiosity that keeps us in the present moment and the kind that might take us out of it—into a world of distraction. We might *think* we're approaching our uncertainty with curiosity by, say, going down an internet rabbit hole to self-diagnose a medical issue, but this behavior can just create a new habit of distracting yourself when you're anxious, Brewer writes. Distractions don't always work, which is why curiosity is the more effective "cure" for being able to live, and love, our questions.[4]

Bust a Myth

The question of how to relate more curiously to our uncertainty is one that Scott Shigeoka also takes up in his book *Seek*. But first, he shares encouraging findings from a meta-analysis of one million participants. Contrary to popular wisdom, we actually become *more* curious as we get older, with the only decrease happening when our overall cognition declines near death. "That widely held idea that kids are more curious than adults—because they ask more questions than us—is actually a myth!" Shigeoka writes. "Whether it's finding a job, listening to other people's stories, or exploring ways to solve problems, we are constantly generating new thoughts and asking small and big questions—curiosity is our partner in life that never leaves our side."[5]

Create an Implementation Intention

Shigeoka suggests practicing exercises that "bring curiosity to the forefront of your conscious mind, orient your thinking around seeking questions rather than answers and tap into your innate power of imagination to prepare you for encounters that rely on your ability to be curious."[6]

One idea he offers is rooted in research on implementation intentions. If you make clear when, where, and how you'll do something, along with what might get in the way of successfully completing your task, you'll be more likely to do it. In this case, Shigeoka suggests creating an implementation intention around curiosity: for example, maybe you see a busy day on the calendar and write yourself a reminder to check in with yourself in the morning and afternoon, asking how you're feeling.[7]

Remember—You Aren't Your Thoughts and Feelings

The attitude of curiosity is about bringing a playful attitude to the thoughts and emotions that arise for you, asking questions about them, and noting your responses. "When you bring in this curious attitude [your thoughts and emotions] are much less likely to have the power they once had over you," Judson Brewer writes. "It becomes obvious that they are just thoughts and sensations in your body. Yes, they may be driving your life for the moment, but they do not constitute who you are."[8]

Questions You Can Ask Yourself to Spark Curiosity and Self-Compassion*

- *What made me stick out as a child or not fit in with the crowd? What makes me "different"?*

---

* These questions were provided by Chrissy. My sincere thanks to her for the contribution.

- *What have been the biggest challenges in my life? What have I learned from them? What have I made them mean?*

- *How is my current life different from my childhood? In what ways am I still obeying "rules" of my childhood that aren't appropriate or necessary now?*

- *What helps me to feel safe? Do I know what safety feels like in my body? How have I worked through difficulty in the past? Who are the people and what are the places and things that help me connect with a sense of safety or "wholeness"?*

- *What aspects of myself could I imagine feeling gratitude or appreciation for, if I really wanted to?*

NOTES

Introduction

1. Frank Hyneman Knight, *Risk, Uncertainty and Profit* (Houghton Mifflin Company, 1921); Sniazhana Sniazhko and Etayankara Muralidharan, "Uncertainty in Decision-Making: A Review of the International Business Literature," *Cogent Business and Management* 6, no. 1 (2019): 2.
2. Raanan Lipshitz and Orna Strauss, "Coping with Uncertainty: A Naturalistic Decision-Making Analysis," *Organizational Behavior and Human Decision Processes* 69, no. 2 (February 1997): 150.
3. Charley M. Wu, Eric Schulz, Timothy J. Pleskac, and Maarten Speekenbrink, "Time Pressure Changes How People Explore and Respond to Uncertainty," *Scientific Reports* 12, no. 1 (2022): 4122.
4. M. A. Addicott, J. M. Pearson, M. M. Sweitzer, D. L. Barack, M. L. Platt, "A Primer on Foraging and the Explore/Exploit Trade-off for Psychiatry Research," *Neuropsychopharmacology* 42 (2017): 1931–39; Barbara Jacquelyn Sahakian and Aleya Aziz Marzuki, "How Uncertainty Can Impair Our Ability to Make Rational Decisions: New Research," *The Conversation*, November 30, 2021, https://theconversation.com/how-uncertainty-can-impair-our-ability-to-make-rational-decisions-new-research-172525.
5. Dan Pilat and Sekoul Krastev, "Why Do We Prefer Options We Know?," The Decision Lab, accessed October 16, 2024, https://thedecisionlab.com/biases/ambiguity-effect.
6. Arne Roets and Alain Van Hiel, "Separating Ability from Need: Clarifying the Dimensional Structure of the Need for Closure Scale," *Personality and Social Psychology Bulletin* 33, no. 2 (2007): 266–80; Donna M. Webster and Arie W. Kruglanski, "Individual Differences in Need for Cognitive Closure," *Journal of Personality and Social Psychology* 67, no. 6 (1994): 1049–62.
7. R. Nicholas Carleton, Gabrielle Desgagné, Rachel Krakauer, and Ryan Y. Hong, "Increasing Intolerance of Uncertainty over Time: The Potential Influence of Increasing Connectivity," *Cognitive Behaviour Therapy* 48, no. 2 (March 2019 [e-published June 8, 2018]): 121–36.

8. Carleton et al., "Increasing Intolerance of Uncertainty over Time," 121.
9. Renee D. Goodwin, Andrea H. Weinberger, June H. Kim, Melody Wu, Sandro Galea, "Trends in Anxiety Among Adults in the United States, 2008–2018: Rapid Increases Among Young Adults," *Journal of Psychiatric Research* 130 (2020): 441–46.
10. Kristeen Cherney, "Effects of Anxiety on the Body: Immune System," Healthline, November 13, 2023, https://www.healthline.com/health/anxiety/effects-on-body#immune.
11. Ida Andersson, Julia Persson, and Petri Kajonius, "Even the Stars Think That I Am Superior: Personality, Intelligence and Belief in Astrology," *Personality and Individual Differences* 187 (March 2022): 111389.
12. Jessica Bursztynsky, "Tired of Swiping Left, Singles Are Turning to New Matchmaking Services for Dates," CNBC, July 4, 2022, https://www.cnbc.com/2022/07/04/tired-of-swiping-left-singles-turn-to-new-matchmaking-services-.html.
13. Alyson Krueger, "What It's Like to Work with a Matchmaker," *New York Times*, February 27, 2021, https://www.nytimes.com/2021/02/27/style/what-its-like-to-work-with-a-matchmaker.html/.
14. International Coaching Federation, "Executive Summary," *2020 ICF Global Coaching Study*, 2020, https://coachingfederation.org/app/uploads/2020/09/FINAL_ICF_GCS2020_ExecutiveSummary.pdf.
15. Elizabeth Weingarten, "Twenty Questions to Ask Instead of 'How Are You Doing Right Now?,'" *Quartz*, April 10, 2020, https://qz.com/work/1836105/20-great-questions-to-ask-instead-of-how-are-you-doing-right-now; Elizabeth Weingarten, "The Gendered Way We've Learned to Ask Questions Is Terrible for Both Men and Women," *Quartz*, March 2, 2016, https://qz.com/628724/the-gendered-way-weve-learned-to-ask-questions-is-terrible-for-both-men-and-women; Elizabeth Weingarten, "Who Asks Questions, and What It Tells Us," *Behavioral Scientist*, June 19, 2019, https://behavioralscientist.org/who-asks-questions-and-what-it-tells-us/.
16. Krista Tippett, *Living the Questions*, *On Being* podcast series, https://onbeing.org/series/living-the-questions/.
17. Rachel Corbett, *You Must Change Your Life: The Story of Rainer Maria Rilke and Auguste Rodin* (W. W. Norton & Company, Inc., 2016), 104.
18. Rainer Maria Rilke and Franz Xaver Kappus, *Letters to a Young Poet: With the Letters to Rilke from the "Young Poet,"* trans. Damion Searls (Liveright, 2020), v–vii.
19. Rilke and Kappus, *Letters to a Young Poet*, 11–13.
20. Rilke and Kappus, *Letters to a Young Poet*, 84–85.
21. Rilke and Kappus, *Letters to a Young Poet*, 22.
22. "Pawpaw: A Big Fruit in Need of Bigger Love," Heritage Conservancy,

November 9, 2023, https://heritageconservancy.org/pawpaw-a-big-fruit-in-need-of-bigger-love/.

23. "Anatomy of a Tree," Arbor Day Foundation, accessed October 16, 2024, https://www.arborday.org/trees/ringstreenatomy.cfm.

Chapter 1: The Charlatans of Certainty

1. Rachel Corbett, *You Must Change Your Life: The Story of Rainer Maria Rilke and Auguste Rodin* (W. W. Norton, 2016), 24–28; Ralph Freedman, *Life of a Poet: Rainer Maria Rilke* (Northwestern Univ. Press, 1996), 60–62.
2. Corbett, *You Must Change Your Life*, 28.
3. Robert Holden, *Happiness Now! Timeless Wisdom for Feeling Good FAST* (Hay House, 2007).
4. Jessica Lamb-Shapiro, *Promise Land: My Journey through America's Self-Help Culture* (Simon & Schuster, 2014).
5. Jessica Lamb-Shapiro, in discussion with the author, January 6, 2023.
6. Lamb-Shapiro, *Promise Land*, 36.
7. Lamb-Shapiro, *Promise Land*, 36.
8. Mark Coleman, "Make Peace with Your Mind: How Mindfulness and Compassion Can Help Free You from the Inner Critic," Esalen, February 17–19, 2023, https://www.esalen.org/workshops/make-peace-with-your-mind-how-mindfulness-and-compassion-can-help-free-you-from-the-inner-critic-21723.
9. Claire Hoffman, *Greetings from Utopia Park: Surviving a Transcendent Childhood* (HarperCollins Publishers, 2016), 148–49.
10. Claire Hoffman, in discussion with the author, January 30, 2023.
11. Amanda Montell, *Cultish: The Language of Fanaticism* (Harper, 2021), 12.
12. Montell, *Cultish*, 23-24; *Fleabag*, season 2, episode 4, "Episode #2.4," directed by Harry Bradbeer, written by Phoebe Waller-Bridge, featuring Phoebe Waller-Bridge, Sian Clifford, and Olivia Colman, aired May 17, 2019, on BBC.
13. Lamb-Shapiro, *Promise Land*, 23.
14. Lamb-Shapiro, discussion.
15. *Merriam-Webster Dictionary*, "seduce," accessed August 6, 2024, https://www.merriam-webster.com/dictionary/seduce.
16. Kurt Lewin, *Principles of Topological Psychology*, trans. Fritz Heider and Grace M. Heider (McGraw-Hill Book Company, Inc., 1936).
17. Archy O. de Berker, Robb B. Rutledge, Christoph Mathys, et al., "Computations of Uncertainty Mediate Acute Stress Responses in Humans," *Nature Communications* 7, no. 1 (2016), 10996; Shawn P. Curley, J. Frank Yates, and Richard A. Abrams, "Psychological Sources of Ambiguity Avoidance,"

Organizational Behavior and Human Decision Processes 38, no. 2 (October 1986): 230–56; David J. Hardisty and Jeffrey Pfeffer, "Intertemporal Uncertainty Avoidance: When the Future Is Uncertain, People Prefer the Present, and When the Present Is Uncertain, People Prefer the Future," *Management Science* 63, no. 2 (2017): 519–27.

18. Achim Peters, Bruce S. McEwen, and Karl Friston, "Uncertainty and Stress: Why It Causes Diseases and How It Is Mastered by the Brain," *Progress in Neurobiology*, no. 156 (September 2017): 164–88.

19. Donna M. Webster and Arie W. Kruglanski, "Individual Differences in Need for Cognitive Closure," *Journal of Personality and Social Psychology* 67, no. 6 (1994): 1049–62.

20. Arne Roets and Alain Van Hiel, "Separating Ability from Need: Clarifying the Dimensional Structure of the Need for Closure Scale," *Personality and Social Psychology Bulletin* 33, no. 2 (2007): 266–80; Webster and Kruglanski, "Individual Differences in Need."

21. Arie Kruglanski, in discussion with the author, November 18, 2022.

22. Arie W. Kruglanski, *The Psychology of Closed Mindedness* (Psychology Press, 2004), 14.

23. Arie W. Kruglanski and Donna M. Webster, "Motivated Closing of the Mind: 'Seizing' and 'Freezing,'" in *The Motivated Mind*, ed. Arie Kruglanski (Routledge, 2018), 60–103.

24. Aaron C. Kay, Jennifer A. Whitson, Danielle Gaucher, and Adam D. Galinsky, "Compensatory Control: Achieving Order Through the Mind, Our Institutions, and the Heavens," *Current Directions in Psychological Science* 18, no. 5 (2009): 264–68.

25. Ryan Perry and Chris G. Sibley, "Seize and Freeze: Openness to Experience Shapes Judgments of Societal Threat," *Journal of Research in Personality* 47, no. 6 (December 2013): 677–86.

26. National Institutes of Health, "About ME/CFS," *Advancing ME/CFS Research*, August 12, 2022, https://www.nih.gov/mecfs/about-mecfs.

27. Paul Weingarten, "We Take It Day by Day," *Chicago Tribune*, October 16, 1992, C1, https://www.chicagotribune.com/1992/10/16/we-take-it-day-by-day/.

28. Gabor Maté with Daniel Maté, *The Myth of Normal: Trauma, Illness, and Healing in a Toxic Culture* (Avery, 2022).

29. Maté, *The Myth of Normal*, 19–20.

30. Maté, *The Myth of Normal*, 21–23.

31. Maté, *The Myth of Normal*, 20.

32. Peter A. Levine, *Healing Trauma: A Pioneering Program to Restore the Wisdom of Your Body* (Sounds True, 2005), 9; Maté, *The Myth of Normal*, 23.

33. Bessel van der Kolk, *The Body Keeps the Score: Brain, Mind, and Body in the Healing of Trauma* (Penguin Books, 2015).
34. Sophie, in discussion with the author, January 17, 2024, and January 26, 2024.
35. Scott Barry Kaufman, *Transcend: The New Science of Self-Actualization* (TarcherPerigree, 2021), 8.
36. Dan Sperber, Fabrice Clément, Christophe Heintz, et al., "Epistemic Vigilance," *Mind and Language*, 25 no. 4 (2010): 359–93.
37. Tania Lombrozo, in discussion with the author, January 3, 2023. For additional information, see: Kerem Oktar and Tania Lombrozo, "Deciding to Be Authentic: Intuition Is Favored over Deliberation When Authenticity Matters," *Cognition* 223 (2022): 105021.
38. Damian A. Stanley, Peter Sokol-Hessner, Mahzarin R. Banaji, and Elizabeth A. Phelps, "Implicit Race Attitudes Predict Trustworthiness Judgments and Economic Trust Decisions," *Proceedings of the National Academy of Sciences* 108, no. 19 (2011): 7710–15; Paul K. Piff, Michael W. Kraus, Stéphane Côté, Bonnie Hayden Cheng, Dacher Keltner, "Having Less, Giving More: The Influence of Social Class on Prosocial Behavior," *Journal of Personality and Social Psychology* 99, no. 5 (2010): 771; Beata Krawczyk-Bryłka, "Trust Triggers and Barriers in Intercultural Teams," *Journal of Intercultural Management* 8, no. 2 (2016): 105–23.
39. Marie Prevost, Mathieu Brodeur, Kristine H. Onishi, Martin Lepage, Ian Gold, "Judging Strangers' Trustworthiness Is Associated with Theory of Mind Skills," *Frontiers in Psychiatry* 6 (2015): 52.
40. Prevost et al., "Judging Strangers' Trustworthiness."
41. Eileen Barker, "The Unification Church," in *America's Alternative Religions*, ed. Timothy Miller (State Univ. of New York Press, 1995), 223–29; Eileen Barker, *The Making of a Moonie: Choice or Brainwashing?* (Blackwell Publishing, 1984); "Unification Church," *Britannica*, last updated July 12, 2024, https://www.britannica.com/topic/Unification-Church.
42. Montell, *Cultish*, 97.
43. Jay Medenwaldt, in discussion with the author, January 12, 2023.
44. Joshua N. Hook, Todd W. Hall, Don E. Davis, Daryl R. Van Tongeren, and Mackenzie Conner, "The Enneagram: A Systematic Review of the Literature and Directions for Future Research," *Journal of Clinical Psychiatry* 77, no. 4 (April 2021): 865–83.
45. Oscar Ichazo, "Letter to the Transpersonal Community," PDFCOFFEE.com, accessed August 4, 2024, https://pdfcoffee.com/letter-to-the-transpersonal-community-pdf-free.html.
46. Aditi Shrikant, "Are You 'The Achiever' or 'The Enthusiast'? Why So Many People—Including Your Boss—Are Obsessed with the Enneagram Test,"

Psychology and Relationships, CNBC.com, March 21, 2023, https://www.cnbc.com/2023/03/21/personality-tests-why-the-enneagram-is-trending-again.html.

47. Enneagram Institute, "About the Enneagram Institute®," accessed August 4, 2024, https://www.enneagraminstitute.com/about.
48. Jay Medenwaldt, "The Enneagram, Science, and Christianity—Part 1," *Jay Medenwaldt* (blog), January 15, 2019, https://jaymedenwaldt.com/the-enneagram-science-and-christianity-part-1/.
49. Medenwaldt, discussion.
50. Hook et al., "The Enneagram."
51. Jay Medenwaldt, "Does the Enneagram Live Up to Its Claims?," paper presented at the Christian Association of Psychological Studies Annual Conference, March 18–19, 2022 (virtual); Jay Medenwaldt, "Eradicating the Enneagram," data blitz presentation, Personality Science Preconference, Society for Personality and Social Psychology Annual Convention, February 7–10, 2024, San Diego, CA; Jay Medenwaldt, "Failed Claims of the Enneagram," paper presented at the Christian Association of Psychological Studies Annual Conference, March 21–23, 2024, Atlanta, GA.
52. Jay Medenwaldt, "The Enneagram, Science, and Christianity—Part 2," *Jay Medenwaldt* (blog), January 17, 2019, https://jaymedenwaldt.com/the-enneagram-science-and-christianity-part-2/.
53. Patrick Healy, "The Fundamental Attribution Error: What It Is and How to Avoid It," *Business Insights*, Harvard Business School Online, June 8, 2017, https://online.hbs.edu/blog/post/the-fundamental-attribution-error.
54. Nick Epley, in discussion with the author, August 2, 2021.
55. Lila Thulin, "The First Personality Test Was Developed During World War I," *Smithsonian Magazine*, September 23, 2019, https://www.smithsonianmag.com/history/first-personality-test-was-developed-during-world-war-i-180973192/.
56. Robert E. Gibby and Michael J. Zickar, "A History of the Early Days of Personality Testing in American Industry: An Obsession with Adjustment," *History of Psychology* 11, no. 3 (August 2008): 164–84.
57. Dan Pilat and Sekoul Krastev, "Why Do We Believe Our Horoscopes? The Barnum Effect, Explained," The Decision Lab, accessed August 3, 2024, https://thedecisionlab.com/biases/barnum-effect.
58. Bertram R. Forer, "The Fallacy of Personal Validation: A Classroom Demonstration of Gullibility," *Journal of Abnormal and Social Psychology* 44, no. 1 (1949): 118–23.
59. Paul E. Meehl, "Wanted—A Good Cook-Book," *American Psychologist* 11, no. 6 (1956): 263–72.
60. Garson O'Toole, "There's a Sucker Born Every Minute," Quote Investigator, April 11, 2014, https://quoteinvestigator.com/2014/04/11/fool-born/.

61. Arie Kruglanski, in discussion with the author, November 18, 2023.
62. Beth Azar, "Positive Psychology Advances, with Growing Pains," *Monitor on Psychology* 42, no. 4 (April 2011): 32.
63. Kruglanski, discussion.
64. Dong Hoon Lee, Leandro F.M. Rezende, Hee-Kyung Joh, et al., "Long-Term Leisure-Time Physical Activity Intensity and All-Cause and Cause-Specific Mortality: A Prospective Cohort of Us Adults," *Circulation* 146, no. 7 (2022): 523–34.

Chapter 2: The Death (and Rebirth) of Patience

1. Ralph Freedman, *Life of a Poet: Rainer Maria Rilke* (Northwestern Univ. Press, 1996), 119–26.
2. Freedman, *Life of a Poet*, 130–33.
3. Freedman, *Life of a Poet*, 138–39, 143.
4. "Amaxophobia (Fear of Driving)," Cleveland Clinic, March 23, 2022, https://my.clevelandclinic.org/health/diseases/22558-amaxophobia-fear-of-driving.
5. E. N. Widayati, N. E. Rakhmawati, and D. Pratama, "The Architectural Structure of Joglo House as the Manifestation of Javanese Local Wisdom," in *Proceedings of 1st Workshop on Environmental Science, Society, and Technology, WESTECH 2018*, December 8, 2018, Medan, Indonesia.
6. Rainer Maria Rilke and Franz Xaver Kappus, *Letters to a Young Poet: With the Letters to Rilke from the "Young Poet,"* trans. Damion Searls (Liveright, 2020), 16.
7. Rilke and Kappus, *Letters to a Young Poet*, 38, 57.
8. Rilke and Kappus, *Letters to a Young Poet*, 61.
9. Sarah A. Schnitker, "Patience and Purpose in the Pursuit of Personal Projects," *Good Thought*, Center for Social Concerns, University of Notre Dame, April 2022, https://socialconcerns.nd.edu/virtues/newsletter-post/patience-and-purpose-in-the-pursuit-of-personal-projects/.
10. David Baily Harned, *Patience: How We Wait upon the World* (Wipf and Stock, 1997).
11. Harned, *Patience*, 1–2.
12. Harned, *Patience*, 4–5.
13. Nicholas Carleton, in discussion with the author, February 3, 2023.
14. For example, see: R. Nicholas Carleton, Gregory P. Krtzig, Shannon Sauer-Zavala, et al. "Study Protocol-The Royal Canadian Mounted Police (RCMP) Study: Protocol for a Prospective Investigation of Mental Health Risk and Resilience Factors," *Health Promotion and Chronic Disease Prevention in Canada: Research, Policy and Practice* 42, no. 8 (2022): 319–33.
15. R. Nicholas Carleton, Gabrielle Desgagné, Rachel Krakauer, and Ryan Y. Hong, "Increasing Intolerance of Uncertainty over Time: The Potential

Influence of Increasing Connectivity," *Cognitive Behaviour Therapy* 48, no. 2 (March 2019 [e-published June 8, 2018]): 121–36.

16. "Scientific Consensus," NASA, https://science.nasa.gov/climate-change/scientific-consensus/; Luke Kemp, Chi Xu, Joanna Depledge, et al., "Climate Endgame: Exploring Catastrophic Climate Change Scenarios," *Proceedings of the National Academy of Sciences* 119, no. 34 (2022): e2108146119.

17. Yang Wu, Lu Wang, Mengjun Tao, et al., "Changing Trends in the Global Burden of Mental Disorders from 1990 to 2019 and Predicted Levels in 25 Years," *Epidemiology and Psychiatric Sciences*, no. 32 (2023): e63.

18. Harned, *Patience*, 5.

19. Sarah Schnitker, in discussion with the author, January 21, 2022.

20. Christopher Peterson and Martin E. P. Seligman, *Character Strengths and Virtues: A Handbook and Classification* vol. 1. (Oxford Univ. Press, 2004); Martin E. P. Seligman and Mihaly Csikszentmihalyi, *Positive Psychology: An Introduction* vol. 55, no. 1. (American Psychological Association, 2000).

21. Sarah A. Schnitker and Justin T. Westbrook, "Do Good Things Come to Those Who Wait?: Patience Interventions to Improve Well-Being," in *The Wiley Blackwell Handbook of Positive Psychological Interventions*, ed. Acacia C. Parks and Stephen M. Schueller (John Wiley & Sons, 2014), 155–67.

22. Schnitker, discussion.

23. Aristotle, *Nicomachean Ethics*, book IV, chap. 5, in *The Basic Works of Aristotle*, trans. W. D. Ross, ed. Richard McKeon (Random House, 1941).

24. Juliette L. Ratchford, Amber R. Cazzell, Elizabeth Wood, Bradley Owens, Robert Quinn, and Sarah A. Schnitker, "The Virtue Counterbalancing Model: An Illustration with Patience and Courage," *Journal of Positive Psychology* 19, no. 3 (2024): 406–18.

25. Sarah A. Schnitker and Robert A. Emmons, "Patience as a Virtue: Religious and Psychological Perspectives," *Research in the Social Scientific Study of Religion*, no. 18 (2007): 177–207.

26. Schnitker, discussion.

27. National Institute of Diabetes and Digestive and Kidney Diseases, "Cyclic Vomiting Syndrome," https://www.niddk.nih.gov/health-information/digestive-diseases/cyclic-vomiting-syndrome.

28. Sarah Schnitker, "An Examination of Patience and Well-Being," *Journal of Positive Psychology* 7, no 4 (2012): 263–80.

29. Schnitker, discussion.

30. Ryan M. Thomas and Sarah A. Schnitker, "Modeling the Effects of Within-Person Characteristic and Goal-Level Attributes on Personal Project Pursuit over Time," *Journal of Research in Personality* 69 (August 2017): 206–17.

31. Jennifer Shubert, Juliette L. Ratchford, Benjamin J. Houltberg, and Sarah A. Schnitker, "Disentangling Character Strengths from Developmental Competencies: The Virtue of Patience and Self-Regulatory Competencies," *Journal of Positive Psychology* 17, no. 2 (2022): 203–9; Sarah A. Schnitker, Diana B. Ro, Joshua D. Foster, Alexis D. Abernethy, Joseph M. Currier, and Charlotte vanOyen Witvlie, "Patient Patients: Increased Patience Associated with Decreased Depressive Symptoms in Psychiatric Treatment," *Journal of Positive Psychology* 15, no. 3 (2020): 300–13; Schnitker and Westbrook, "Do Good Things Come to Those Who Wait?"

32. Harned, *Patience*, 101.

33. Schnitker, discussion.

34. Ratchford et al., "The Virtue Counterbalancing Model."

35. Tara Isabella Burton, "The Waco Tragedy, Explained," *Vox*, March 23, 2023, https://www.vox.com/2018/4/19/17246732/waco-tragedy-explained-david-koresh-mount-carmel-branch-davidian-cult-30-year-anniversary.

36. Ryan B. Martinez, "The Many Murals of Waco," *Baylor Line*, March 15, 2024, https://baylorline.com/the-many-murals-of-waco/.

37. Schnitker, discussion.

38. Juliette Ratchford and Merve Balkaya-Ince, in discussion with the author, January 13, 2023.

39. Juliette Ratchford, in discussion with the author, January 13, 2023.

40. Elizabeth Bounds and Jay Medenwaldt, in discussion with the author, January 12, 2023.

41. Britton W. Brewer and Albert J. Petitpas, "Athletic Identity Foreclosure," *Current Opinion in Psychology* 16 (2017): 118–22.

42. Schnitker, discussion.

43. Schnitker, "An Examination of Patience and Well-Being."

44. Schnitker, discussion.

45. "A Definition of Motivational Interviewing," UMass Amherst, accessed October 16, 2024, https://www.umass.edu/studentlife/sites/default/files/documents/pdf/Motivational_Interviewing_Definition_Principles_Approach.pdf.

46. Gabriele Oettingen and Peter M. Gollwitzer, "Strategies of Setting and Implementing Goals: Mental Contrasting and Implementation Intentions," in *Social Psychological Foundations of Clinical Psychology*, ed. James E. Maddux (Guilford, 2010), 114–35.

47. Guoxia Wang, Yi Wang, and Xiaosong Gai, "A Meta-Analysis of the Effects of Mental Contrasting with Implementation Intentions on Goal Attainment," *Frontiers in Psychology* 12 (2021): 565202.

48. Rilke and Kappus, *Letters to a Young Poet*, 54–55.

49. Office of the Surgeon General, *Our Epidemic of Loneliness and Isolation: The U.S. Surgeon General's Advisory on the Healing Effects of Social Connection and Community* (US Department of Health and Human Services, 2023).
50. Kristin Kerns-D'Amore, Joey Marshall, and Brian McKenzie, "Pandemic Did Not Disrupt Decline in Rate of People Moving," United States Census Bureau, March 7, 2022, https://www.census.gov/library/stories/2022/03/united-states-migration-continued-decline-from-2020-to-2021.html.
51. Office of the Surgeon General, *Our Epidemic of Loneliness*.

Chapter 3: Seeking the Right Sources to Ask

1. Ralph Freedman, *Life of a Poet: Rainer Maria Rilke* (Northwestern Univ. Press, 1996), 152–54.
2. Freedman, *Life of a Poet*, 157.
3. Freedman, *Life of a Poet*, 155, 159–60.
4. Sabrina Stierwalt, "Is Astrology Real? Here's What Science Says," *Scientific American*, June 25, 2020, https://www.scientificamerican.com/article/is-astrology-real-heres-what-science-says/.
5. Sydney Page, "Young People Are Flocking to Astrology. But It Comes with Risks," *Washington Post*, June 13, 2023, https://www.washingtonpost.com/lifestyle/2023/06/13/astrology-millennials-gen-z-science/.
6. Jessica Lanyadoo, in discussion with author, August 18, 2023.
7. Diana Rose Harper, "How Do You Practice Responsible Astrology?," *Wired*, January 5, 2022, https://www.wired.com/story/astrology-prediction-ethics/.
8. For example, see: John M. Gottman and Nan Silver, *The Seven Principles for Making Marriage Work: A Practical Guide from the Country's Foremost Relationship Expert* (Three Rivers Press, 1999); Esther Perel, *Mating in Captivity: Unlocking Erotic Intelligence* (HarperCollins Publishers, 2007); Esther Perel, *Where Should We Begin* podcast, Esther Perel Global Media, https://podcasts.apple.com/us/podcast/where-should-we-begin-with-esther-perel/id1237931798.
9. Mateo, in discussion with author, February 25, 2023.
10. National Academies of Sciences, Engineering, and Medicine, *How People Learn II: Learners, Contexts, and Cultures* (Washington, DC: The National Academies Press, 2018).
11. Raquel Lozano-Blasco, Alberto Quilez Robres, and Alberto Soto Sánchez, "Internet Addiction in Young Adults: A Meta-Analysis and Systematic Review," *Computers in Human Behavior* 130 (2022): 107201.
12. Andrew Perrin and Sara Atske, "About Three-in-Ten US Adults Say They Are 'Almost Constantly' Online," *Short Reads*, Pew Research Center, March 26, 2021, https://www.pewresearch.org/short-reads/2021/03/26/about-three-in-ten-u-s-adults-say-they-are-almost-constantly-online/.

13. Kenji Kobayashi and Ming Hsu, "Common Neural Code for Reward and Information Value," *Proceedings of the National Academy of Sciences* 116, no. 26 (2019): 13061–66.

14. Laura Counts, "How Information Is Like Snacks, Money, and Drugs—to Your Brain," June 19, 2019, Haas School of Business, University of California, Berkeley, https://newsroom.haas.berkeley.edu/how-information-is-like-snacks-money-and-drugs-to-your-brain/.

15. Ciera Kirkpatrick, in discussion with the author, March 3, 2023, and June 3, 2024.

16. Edison Research, "Moms and Media 2020," May 7, 2020, https://www.edisonresearch.com/moms-and-media-2020/.

17. Edison Research, "Moms and Media 2023," May 11, 2023, https://www.edisonresearch.com/moms-and-media-2023/.

18. Ciera E. Kirkpatrick and Sungkyoung Lee, "Comparisons to Picture-Perfect Motherhood: How Instagram's Idealized Portrayals of Motherhood Affect New Mothers' Well-Being," *Computers in Human Behavior* 137 (December 2022): 107417.

19. Rachel Y. Moon, Anita Mathews, Rosalind Oden, and Rebecca Carlin, "Mothers' Perceptions of the Internet and Social Media as Sources of Parenting and Health Information: Qualitative Study," *Journal of Medical Internet Research* 21, no. 7 (2019): e14289.

20. Jessica Grose, in discussion with the author, March 7, 2023.

21. Jessica Grose, *Screaming on the Inside: The Unsustainability of American Motherhood* (HarperCollins Publishers, 2022).

22. Caitlyn Collins, *Making Motherhood Work: How Women Manage Careers and Caregiving* (Princeton University Press, 2019).

23. Caitlyn Collins, "Why American Moms Can't Get Enough Expert Parenting Advice," *The Atlantic*, May 12, 2019, https://www.theatlantic.com/family/archive/2019/05/american-parents-obsession-expert-advice/589132/.

24. *Her*, directed by Spike Jonze (Warner Bros. Pictures, 2013).

25. Joseph Weizenbaum, "ELIZA—A Computer Program for the Study of Natural Language Communication Between Man and Machine," *Communications of the ACM* 9, no. 1 (1966): 36–45.

26. Joseph Weizenbaum with Gunna Wendt, *Islands in the Cyberstream: Seeking Havens of Reason in a Programmed Society*, trans. Benjamin Fasching-Gray (Litwin Books, 2015), 87.

27. Weizenbaum, *Islands in the Cyberstream*, 90.

28. Lucy Yao and Rian Kabir, *Person-Centered Therapy (Rogerian Therapy)* (StatPearls Publishing, 2023).

29. Weizenbaum, *Islands in the Cyberstream*, 90–91.

30. Weizenbaum, *Islands in the Cyberstream*, 90.
31. Joseph Weizenbaum, "On the Impact of the Computer on Society: How Does One Insult a Machine?," *Socio-Anthropologie* 47 (2023): 193–206.
32. Weizenbaum, *Islands in the Cyberstream*, 2, 21–23.
33. Joseph Weizenbaum, *Computer Power and Human Reason: From Judgment to Calculation* (W. H. Freeman, 1976), 223.
34. Joseph Weizenbaum, "Professor Emeritus of Computer Science, 85," *MIT News*, March 10, 2008, https://news.mit.edu/2008/obit-weizenbaum-0310.
35. Henry Kissinger, Eric Schmidt, and Daniel Huttenlocher, "ChatGPT Heralds an Intellectual Revolution," *Wall Street Journal*, February 24, 2023, https://www.wsj.com/articles/chatgpt-heralds-an-intellectual-revolution-enlightenment-artificial-intelligence-homo-technicus-technology-cognition-morality-philosophy-774331c6.
36. Xuan Zhao and Nicholas Epley, "Surprisingly Happy to Have Helped: Underestimating Prosociality Creates a Misplaced Barrier to Asking for Help," *Psychological Science* 33, no. 10 (September 6, 2022).
37. Xuan Zhao, in discussion with the author, March 8, 2023.
38. Jessica Lanyadoo, in discussion with author, August 18, 2023.
39. Mateo, discussion.

Chapter 4: What Does a Life of Loving Questions Look Like?

1. Rachel Corbett, *You Must Change Your Life: The Story of Rainer Maria Rilke and Auguste Rodin* (W. W. Norton, 2016), 114; Ralph Freedman, *Life of a Poet: Rainer Maria Rilke* (Northwestern Univ. Press, 1996), 197–98.
2. Corbett, *You Must Change Your Life*, 104.
3. Corbett, *You Must Change Your Life*, 110.
4. Rainer Maria Rilke, *Letters to a Young Poet*, revised ed., trans. M. D. Herter Norton (New York: W. W. Norton, 1962), 27.
5. Rachel Corbett, in discussion with the author, July 22, 2021.
6. Parker Palmer, in discussion with the author, multiple dates between June 25, 2021, and May 24, 2024.
7. "Parker J. Palmer," Center for Courage and Renewal, accessed August 4, 2024, https://couragerenewal.org/parker-j-palmer/.
8. Richard Schwartz, *No Bad Parts: Healing Trauma and Restoring Wholeness with the Internal Family Systems Model* (Sounds True Adult, 2021), 1.
9. Schwartz, *No Bad Parts*, 28.
10. Schwartz, *No Bad Parts*, 28–29.
11. Schwartz, *No Bad Parts*, 29.

12. Martha Beck, *The Way of Integrity: Finding the Path to Your True Self* (Open Field, 2021).
13. Rainer Maria Rilke, *Letters on Life: New Prose Translations*, trans. Ulrich Baer (Random House, 2017), 41.
14. Rilke, *Letters on Life*, 41.
15. Parker Palmer, in discussion with the author, May 17, 2023.
16. Carl Gustav Jung, *Collected Works of CG Jung: Two Essays in Analytical Psychology*, 2nd ed. (Princeton Univ. Press, 1966).
17. Parker J. Palmer, *The Courage to Teach: Exploring the Inner Landscape of a Teacher's Life* (John Wiley & Sons, 2017).
18. Parker J. Palmer, *Let Your Life Speak: Listening for the Voice of Vocation* (John Wiley & Sons, 2000).
19. Sharon and Parker Palmer, in discussion with the author, May 17, 2023.
20. Cameron Crowe, dir., *Jerry Maguire* (Gracie Films, 1996).
21. Rilke, *Letters on Life*, 41.
22. Marianna Jamadi, in discussion with the author, May 24, 2021.
23. Icatsstaci with Pericles Rosa and Nic Chapman, "Cremation Ghats of Varanasi," Atlas Obscura, accessed August 3, 2024, https://www.atlasobscura.com/places/cremation-ghats-of-varanasi.
24. Kunokaiku, accessed October 16, 2023, https://www.kunokaiku.com/.
25. Sheau Yun Melody Ku, "Examining the Relationships Among Uncertainty, Control, Emotionality, and Information Behavior in Patient Experience During In-Vitro Fertilization Treatments" (PhD diss., University of Michigan, 2020); "Fact Sheet: In Vitro Fertilization (IVF) Use Across the United States," U.S. Department of Health and Human Services, March 13, 2024, https://www.hhs.gov/about/news/2024/03/13/fact-sheet-in-vitro-fertilization-ivf-use-across-united-states.html.
26. Sue Saunders, *Maybe Baby: Navigating the Emotional Journey Through Assisted Fertility* (Penguin Random House New Zealand, 2021).
27. Marianna Jamadi, email with the author, December 20, 2023.
28. "Thomas Rockwell: Chief Creative Officer," Exploratorium, accessed August 4, 2024, https://www.exploratorium.edu/about/senior-management/thomas-rockwell.
29. Tom Rockwell, in discussion with the author, May 5, 2023.
30. Deborah Solomon, *American Mirror: The Life and Art of Norman Rockwell* (Farrar, Straus and Giroux, 2013), 3–5.
31. Norman Rockwell, *Freedom from Want*, painting, *Saturday Evening Post*, March 6, 1943; Norman Rockwell, *First Haircut*, painting, American Family Life series for Mass Mutual Insurance Company, 1959; Norman Rockwell,

Boy and Girl Gazing at the Moon (Little Spooners or *Sunset)*, painting, *Saturday Evening Post*, April 24, 1926.

32. Solomon, *American Mirror*, 6, 10.
33. Norman Rockwell, *Portrait of Dwight D. Eisenhower*, painting, *Saturday Evening Post*, October 11, 1952; Norman Rockwell, *Portrait of John F. Kennedy*, painting, *Saturday Evening Post*, October 29, 1960; Norman Rockwell, *Portrait of Lyndon B. Johnson*, painting, 1968; Norman Rockwell, *Portrait of Richard Nixon*, painting, *Saturday Evening Post*, November 5, 1960.
34. Solomon, *American Mirror*, 9.
35. Solomon, *American Mirror*, 134–5, 226–37, 306.
36. Erik H. Erikson, *Childhood and Society* (W. W. Norton, 1950).
37. Solomon, *American Mirror*, 322.
38. Solomon, *American Mirror*, 5.
39. Tom Carson, "The Awakening of Norman Rockwell," *Vox*, February 26, 2020, https://www.vox.com/the-highlight/2020/2/19/21052356/norman-rockwell-the-problem-we-all-live-with-saturday-evening-post.
40. Interview with Mary and Norman Rockwell by Edward R. Murrow, *Person to Person*, aired February 1959 on CBS, https://www.youtube.com/watch?v=SAnrM4wO1qQ.
41. Solomon, *American Mirror*, 5.
42. "Rockwell, Norman," Encyclopedia.com, https://www.encyclopedia.com/people/literature-and-arts/american-art-biographies/norman-rockwell.
43. Solomon, *American Mirror*, 305, 312.
44. Solomon, *American Mirror*, 322.
45. "Peter Rockwell," Norman Rockwell Museum, accessed August 4, 2024, https://www.illustrationhistory.org/artists/peter-rockwell; "In Memoriam: Peter Barstow Rockwell (1936–2020)," Norman Rockwell Museum, https://www.nrm.org/2020/02/peter-barstow-rockwell/.
46. Tom Rockwell, in discussion with the author, various dates between May 5, 2023, and May 24, 2024.
47. Nicholas Stanley-Price, "Peter Rockwell (1936–2020)," International Centre for the Study of the Preservation and Restoration of Cultural Property, March 3, 2020, https://www.iccrom.org/news/peter-rockwell-1936-2020; "Interactive Gargoyle Map," Washington National Cathedral, accessed August 4, 2024, https://www.give2wnc.org/gargoyle/indexA.php.
48. Bernardo Bertolucci, dir. *Luna* (New World Pictures, 1979).
49. Tom Rockwell, in discussion with the author, various dates between May 5, 2023, and May 24, 2024.
50. Rainer Maria Rilke, *Letters on Life: New Prose Translations*, trans. Ulrich Baer (Random House, 2017), 41.

Chapter 5: How Do We Fall in Love with Questions?

1. Rainer Maria Rilke, *Letters on Life: New Prose Translations*, trans. Ulrich Baer (Random House, 2017), vii–ix.
2. "New Year Letter, January 1, 1940, Part I," in *The Collected Poetry of W. H. Auden* (Random House, 1945), 271.
3. Rilke, *Letters on Life*, xxiii.
4. Rainer Maria Rilke and Maurice Betz, *Rilke in Paris*, ed. and trans. Will Stone (Pushkin Press, 2019), 77.
5. Rilke, *Letters on Life*, xii–xxi.
6. Rainer Maria Rilke and Franz Xaver Kappus, *Letters to a Young Poet: With the Letters to Rilke from the "Young Poet,"* trans. Damion Searls (Liveright, 2020), xi.
7. Allen W. Johnson and Timothy K. Earle, *The Evolution of Human Societies: From Foraging Group to Agrarian State* (Stanford Univ. Press, 2000).
8. Stevan E. Hobfoll, *Stress, Culture, and Community: The Psychology and Philosophy of Stress* (Springer, 2004); Casper ter Kuile, *The Power of Ritual: Turning Everyday Activities into Soulful Practices* (HarperOne, 2020), xi.
9. Casper ter Kuile, in discussion with the author, July 8, 2023.
10. "A Program for Deep Connection," Nearness, accessed October 16, 2024, https://www.nearness.coop/.
11. Ella Fitzgerald and Louis Armstrong, "The Nearness of You," from *The Complete Ella and Louis on Verve*, released May 20, 1997.
12. Alec Gewirtz, in discussion with the author, July 18, 2023.
13. Alec Gewirtz, discussion.
14. L'Arche, "About L'Arche," accessed August 4, 2024, https://www.larche.org/about-larche/.
15. Rachel Boughton, in discussion with the author, July 24, 2023.
16. Katie Avis-Riordan, "Ginkgo Biloba: The Tree That Outlived the Dinosaurs," Royal Botanic Gardens: Kew Gardens, May 5, 2020, https://www.kew.org/read-and-watch/ginkgo-biloba-maidenhair-tree-kew-gardens.
17. David Whyte, "Sometimes," in *Everything Is Waiting for You* (Many Rivers Press, 2003).
18. Rilke, *Letters on Life*, 31.
19. Megan Rundel, in discussion with the author, September 11, 2023.
20. Estelle Frankel, *The Wisdom of Not Knowing: Discovering a Life of Wonder by Embracing Uncertainty* (Shambhala, 2017), 45.
21. Joan Sutherland, *Through Forests of Every Color: Awakening with Koans* (Shambhala, 2022), 19–20.
22. Frankel, *Wisdom of Not Knowing*, 45–56.

23. Sutherland, *Through Forests of Every Color*, 5–6, 31–35.
24. Sutherland, *Through Forests of Every Color*, 32.
25. Sutherland, *Through Forests of Every Color*, 6.
26. Sutherland, *Through Forests of Every Color*, 40.
27. John D. Durand, "The Population Statistics of China, AD 2–1953," *Population Studies* 13, no. 3 (1960): 209–56; Peter D. Hershock, *Chan Buddhism* (Univ. of Hawai'i Press, 2005), 32–33.
28. Sutherland, *Through Forests of Every Color*, 30–31.
29. John Tarrant, *Bring Me the Rhinoceros—and Other Zen Koans That Will Save Your Life*, reprint ed. (Shambhala, 2008), 2.
30. Hongzhi Zhengjue and Wansong Xingxiu, "Dizang's Most Intimate," in *The Book of Serenity*, trans. Joan Sutherland and John Tarrant, https://joansutherlanddharmaworks.org/cgi-bin/dharma/jump.cgi?ID=201;view=PDF1, 9.
31. Wiebke Bleidorn and Carolin Ködding, "The Divided Self and Psychological (Mal) Adjustment—A Meta-Analytic Review," *Journal of Research in Personality* 47, no. 5 (2013): 547–52.
32. Wumen Huikai, "Qingshui, Alone and Destitute," in *The Gateless Gateway*, ed. Joan Sutherland, based on trans. Shibayama Zenkei, https://joansutherlanddharmaworks.org/cgi-bin/dharma/jump.cgi?ID=202;view=PDF1, 13.
33. Rundel, discussion.
34. Rundel, discussion.
35. "Zazen Instructions: How to Meditate," Zen Mountain Monastery, accessed July 31, 2024, https://zmm.org/teachings-and-training/meditation-instructions/.
36. Joan Sutherland, in discussion with the author, various dates between July 28, 2023, and March 9, 2024.
37. Sutherland, discussion.
38. Sutherland, discussion.
39. "Mount Rainier National Park, Washington," National Park Service, accessed August 4, 2024, https://www.nps.gov/mora/planyourvisit/climbing.htm; CBS/AP, "Man Falls to Death While Climbing Mount Rainier with Friends," CBS News, August 26, 2022, https://www.cbsnews.com/news/chun-hui-zhang-falls-to-death-climbing-mount-rainier/.
40. "Annual Climbing Statistics," Mount Rainier, National Park Washington, National Park Service, https://www.nps.gov/mora/planyourvisit/annual-climbing-statistics.htm; "Mount Rainier Incidents," National Park Service, accessed August 4, 2024, http://npshistory.com/morningreport/incidents/mora.htm.
41. CNNWire, "Heat Wave: Portland and Seattle See Record Highs Soar Past

100 Degrees," Eyewitness News, ABC7, June 28, 2021, https://abc7.com/seattle-heat-wave-what-is-causing-the-in-pacific-northwest-portland-canadian/10839306/.

42. Charlotte Austin, "One of the Deadliest Days on Mt. Rainier: One Year Later," *Seattle Magazine*, April 29, 2015, https://seattlemag.com/one-deadliest-days-mt-rainier-one-year-later/.

43. "Maps," Mount Rainier, National Park Washington, National Park Service, accessed August 4, 2024, https://www.nps.gov/mora/planyourvisit/maps.htm.

44. Sutherland, discussion.

45. "Dancing Mountains (MK5)," Pacific Zen Institute, PZI Miscellaneous Koans, https://www.pacificzen.org/library/koan/dancing-mountains-mk5/; "Make the Mountains Dance," in *Gates: Miscellaneous Koans*, trans. Joan Sutherland, 17, https://joansutherlanddharmaworks.org/cgi-bin/dharma/jump.cgi?ID=199;view=PDF1; The Hidden Japan, "Yamabushi Mountain Monks," accessed August 2, 2024, https://thehiddenjapan.com/yamabushimountainmonks/.

46. "First Gate," in *Gates: Miscellaneous Koans*, trans. Joan Sutherland, 58, https://joansutherlanddharmaworks.org/cgi-bin/dharma/jump.cgi?ID=199;view=PDF1.

47. Leslie Feinberg, *Stone Butch Blues* (Firebrand Books, 1993).

48. Oliver, in discussion with the author, various dates between April 23, 2023, and May 20, 2024.

49. Jody L. Herman, Andrew R. Flores, Kathryn K. O'Neill, *How Many Adults and Youth Identify as Transgender in the United States?*, School of Law, Williams Institute, UCLA, June 2022, https://williamsinstitute.law.ucla.edu/wp-content/uploads/Trans-Pop-Update-Jun-2022.pdf; Sam Winter, Milton Diamond, Jamison Green, et al., "Transgender People: Health at the Margins of Society," *Lancet* 388, no. 10042 (2016): 390–400; Herman et al., *Identify as Transgender*, 1.

Chapter 6: Converting Questions into Freedom

1. Lesley Chamberlain, *Rilke: The Last Inward Man* (Pushkin Press, 2022).

2. Chamberlain, *Rilke*, 17–18.

3. Rainer Maria Rilke and Franz Xaver Kappus, *Letters to a Young Poet: With the Letters to Rilke from the "Young Poet,"* trans. Damion Searls (Liveright, 2020), 54–55.

4. Anita Barrows and Joanna Macy, trans. and eds., *In Praise of Mortality: Selections from Rainer Maria Rilke's "Duino Elegies" and "Sonnets to Orpheus"* (Echo Point, 2022), 23.

5. Barbara Schoen, in discussion with the author, July 25, 2023, and October 26, 2023.

6. Schoen, discussion.
7. Beatrice A. Wright, *Physical Disability: A Psychological Approach* (Harper & Row; 1960). See also, Dana S. Dunn and Timothy R. Elliott, "Revisiting a Constructive Classic: Wright's Physical Disability: A Psychosocial Approach," *Rehabilitation Psychology* 50, no. 2 (2005): 183.
8. Rashedur Chowdhury, "Misrepresentation of Marginalized Groups: A Critique of Epistemic Neocolonialism," *Journal of Business Ethics* 186, no. 3 (2023): 553–70.
9. Miguel, in discussion with the author, various dates between August 22, 2023, and October 16, 2023.
10. US Citizenship and Immigration Services, "Consideration of Deferred Action for Childhood Arrivals (DACA)," April 8, 2024, https://www.uscis.gov/DACA.
11. "DACA Litigation Information and Frequently Asked Questions," U.S. Citizenship and Immigration Services, September 18, 2023, https://www.uscis.gov/humanitarian/consideration-of-deferred-action-for-childhood-arrivals-daca/daca-litigation-information-and-frequently-asked-questions.
12. Hispanic Association of Colleges and Universities, "Hispanic-Serving Institution Definitions," accessed August 1, 2024, https://www.hacu.net/hacu/HSI_Definition1.asp.
13. Jose Antonio Vargas, *Dear America: Notes of an Undocumented Citizen* (Dey Street, 2018), 71.
14. Jose Antonio Vargas, "My Life as an Undocumented Immigrant," *New York Times Magazine*, June 22, 2011, https://www.nytimes.com/2011/06/26/magazine/my-life-as-an-undocumented-immigrant.html.
15. Carl Davis, Marco Guzman, and Emma Sifre, *Tax Payments by Undocumented Immigrants* (Institute on Taxation and Economic Policy, 2024), https://sfo2.digitaloceanspaces.com/itep/ITEP-Tax-Payments-by-Undocumented-Immigrants-2024.pdf.
16. "Tax Payments by Undocumented Immigrants," Institution on Taxation and Economic Policy, July 30, 2024, https://itep.org/undocumented-immigrants-taxes-2024/.
17. Vargas, *Dear America*, 126.
18. Miguel, discussion.
19. "Travel for DACA Applicants (Advance Parole)," Immigrant Legal Resource Center, 2015, https://www.ilrc.org/sites/default/files/documents/advance_parole_guide.pdf.
20. Joseph Stramondo, in discussion with the author, July 28, 2023, and October 19, 2023.
21. Rosemarie Garland-Thomson, "Misfits: A Feminist Materialist Disability Concept," *Hypatia* 26, no. 3 (2011): 591–609.

22. Joseph A. Stramondo, "Why Bioethics Needs a Disability Moral Psychology," *Hastings Center Report* 46, no. 3 (2016): 22–30.
23. Catriona Mackenzie and Natalie Stoljar, eds., *Relational Autonomy: Feminist Perspectives on Autonomy, Agency, and the Social Self* (Oxford Univ. Press, 2000).
24. Tom Shakespeare, *Disability Rights and Wrongs* (Routledge, 2006).
25. Ae-Kyung Jung and Karen M. O'Brien, "The Profound Influence of Unpaid Work on Women's Lives: An Overview and Future Directions," *Journal of Career Development* 46, no. 2 (2019): 184–200; Natalia Reich-Stiebert, Laura Froehlich, and Jan-Bennet Voltmer, "Gendered Mental Labor: A Systematic Literature Review on the Cognitive Dimension of Unpaid Work Within the Household and Childcare," *Sex Roles* 88, no. 11 (2023): 475–94.
26. Schoen, discussion.
27. Cal Fussman, "Robert Caro: What I've Learned," *Esquire*, December 16, 2009, https://www.esquire.com/entertainment/interviews/a6826/robert-caro-0110/.
28. Barrows and Macy, *In Praise of Mortality*, 23.

Bridge to Part Three: Making Your Questions Map

1. Nick Kabrél, in discussion with the author, December 12, 2023.
2. Carl G. Jung, *Aion: Researches into the Phenomenology of the Self*, trans. R. F. C. Hull (Princeton Univ. Press, 1978).
3. Mykyta Kabrel, Kadi Tulver, and Jaan Aru, "The Journey Within: Mental Navigation as a Novel Framework for Understanding Psychotherapeutic Transformation," *BMC Psychiatry* 24, no. 1 (February 1, 2024): 91.
4. Russell A. Epstein, Eva Zita Patai, Joshua B. Julian, and Hugo J. Spiers, "The Cognitive Map in Humans: Spatial Navigation and Beyond," *Nature Neuroscience* 20, no. 11 (2017): 1504–13.
5. Nick Kabrél, "Boost Your Self-Understanding with a Navigational Approach," *Psyche*, December 5, 2023, https://psyche.co/ideas/boost-your-self-understanding-with-a-navigational-approach.

Chapter 7: Reviving Curiosity When Questions Are Painful

1. Ralph Freedman, *Life of a Poet: Rainer Maria Rilke* (Northwestern Univ. Press, 1996), 436.
2. Freedman, *Life of a Poet*, 455.
3. Freedman, *Life of a Poet*, 455–57.
4. Freedman, *Life of a Poet*, 300, 459.
5. Freedman, *Life of a Poet*, 461.
6. Freedman, *Life of a Poet*, 461.
7. As of this writing, the FDA had ruled to reject MDMA for the purposes of

assisted psychotherapy, though researchers and advocates had plans to appeal the decision and voiced confidence that this would not be the end of the road for the work to legalize therapeutic psychedelic use. Deidre McPhillips and Carma Hassan, "FDA Rejection of MDMA-Assisted Therapy Rattles Advocates but Doesn't Spell End of Psychedelics in Medicine, Experts Say," CNN, August 13, 2024, https://www.cnn.com/2024/08/13/health/mdma-ptsd-studies-retracted-whats-next/index.html; Will Stone, "FDA Gives Thumbs Down to MDMA for Now, Demanding Further Research," NPR, August 9, 2024, https://www.npr.org/sections/shots-health-news/2024/08/09/nx-s1-5068634/mdma-therapy-fda-decision-ptsd-psychedelic-treatment.

8. "MDMA (Ecstasy/Molly)," National Institute on Drug Abuse, accessed August 4, 2024, https://nida.nih.gov/research-topics/mdma-ecstasy-molly.

9. Perry Zurn and Dani S. Bassett, *Curious Minds: The Power of Connection* (MIT Press, 2022), 3.

10. Zurn and Bassett, *Curious Minds*, 4.

11. Nur Hani Zainal and Michelle G. Newman, "Curiosity Helps: Growth in Need for Cognition Bidirectionally Predicts Future Reduction in Anxiety and Depression Symptoms Across Ten Years," *Journal of Affective Disorders* 296 (January 1, 2022): 642–52.

12. Nur Hani Zainal, in discussion with the author, May 22, 2023.

13. Steven Glass, "Truth and D.A.R.E.," *Rolling Stone*, March 5, 1998, https://www.pbs.org/wgbh/pages/frontline/shows/dope/dare/truth.html; "War on Drugs," *Britannica*, last updated May 27, 2024, https://www.britannica.com/topic/war-on-drugs.

14. Tess Borden, *Every 25 Seconds: The Human Toll of Criminalizing Drug Use in the United States* (Human Rights Watch, 2016).

15. Graham Boyd, "The Drug War Is the New Jim Crow," *NACLA Report on the Americas*, July/August 2001, https://www.aclu.org/documents/drug-war-new-jim-crow.

16. Michael Pollan, *How to Change Your Mind: What the New Science of Psychedelics Teaches Us About Consciousness, Dying, Addiction, Depression, and Transcendence* (Penguin Press, 2018); "MDMA," Multidisciplinary Association for Psychedelic Studies, accessed August 2, 2024, https://maps.org/mdma/; NIH Catalyst Staff, "The SIG Beat: Psychedelic Medicine," *NIH Catalyst* 31, no. 2 (March/April 2023), https://irp.nih.gov/catalyst/31/2/the-sig-beat-psychedelic-medicine; "Psychedelic Research and Psilocybin Therapy," Psychiatry and Behavioral Sciences, Johns Hopkins Medicine, accessed August 2, 2024, https://www.hopkinsmedicine.org/psychiatry/research/psychedelics-research.

17. Dakota Sicignano, Kimberly Snow-Caroti, Adrian V. Hernandez, and C. Michael White, "The Impact of Psychedelic Drugs on Anxiety and

Depression in Advanced Cancer or Other Life-Threatening Disease: A Systematic Review with Meta-Analysis," *American Journal of Clinical Oncology* 46, no. 6 (June 2023): 236–45.

18. Alan K. Davis, Frederick S. Barrett, Darrick G. May, et al., "Effects of Psilocybin-Assisted Therapy on Major Depressive Disorder: A Randomized Clinical Trial," *JAMA Psychiatry* 78, no. 5 (2021): 481–89.

19. Alicia L. Danforth, Charles S. Grob, Christopher Struble, et al., "Reduction in Social Anxiety After MDMA-Assisted Psychotherapy with Autistic Adults: A Randomized, Double-Blind, Placebo-Controlled Pilot Study," *Psychopharmacology* 235, no. 11 (2018): 3137–48.

20. Jules Evans, Oliver C. Robinson, Eirini Ketzitzidou, et al., "Extended Difficulties Following the Use of Psychedelic Drugs: A Mixed Methods Study," *PLOS ONE* 18, no. 10 (2023): e0293349.

21. Imran Khan, in discussion with the author, December 3, 2023.

22. Natalie Lyla Ginsberg, in discussion with the author, October 23, 2023.

23. Judith Kerr, *When Hitler Stole Pink Rabbit* (William Collins Sons, 1971).

24. Eric C. Anderson, R. Nicholas Carleton, Michael Diefenbach, and Paul K. J. Han, "The Relationship Between Uncertainty and Affect," *Frontiers in Psychology* 10 (2019): 2504.

25. Bridget Huber, "What Do We Know About the Risks of Psychedelics?," MichaelPollan.com, accessed August 4, 2024, https://michaelpollan.com/psychedelics-risk-today/.

26. Chrissy, in discussion with the author, various dates between May 19, 2023, and September 15, 2023.

27. See also Philip Goff, "Science as We Know It Can't Explain Consciousness—but a Revolution Is Coming," *The Conversation*, November 1, 2019, https://theconversation.com/science-as-we-know-it-cant-explain-consciousness-but-a-revolution-is-coming-126143.

28. R. L. Gregory, *Eye and Brain: The Psychology of Seeing* (Weidenfeld and Nicolson, 1972), 7.

29. Andy Clark, *Surfing Uncertainty: Prediction, Action, and the Embodied Mind* (Oxford Univ. Press, 2015); Ralph Lewis, "The Brain as a Prediction Machine: The Key to Consciousness?," *Psychology Today*, January 1, 2022, https://www.psychologytoday.com/us/blog/finding-purpose/202201/the-brain-prediction-machine-the-key-consciousness.

30. Jakob Hohwy, *The Predictive Mind* (Oxford Univ. Press, 2013).

31. Khan, discussion.

32. Manoj K. Doss, Michal Považan, Monica D. Rosenberg, et al., "Psilocybin Therapy Increases Cognitive and Neural Flexibility in Patients with Major Depressive Disorder," *Translational Psychiatry* 11, no. 1 (2021): 574.

33. Romain Nardou, Eastman M. Lewis, Rebecca Rothhaas, et al., "Oxytocin-Dependent Reopening of a Social Reward Learning Critical Period with MDMA," *Nature* 569, no. 7754 (2019): 116–20; Romain Nardou, Edward Sawyer, Young Jun Song, et al., "Psychedelics Reopen the Social Reward Learning Critical Period," *Nature* 618, no. 7966 (2023): 790–98.

34. Eric R. Kandel, James H. Schwartz, Thomas M. Jessell, Steven Siegelbaum, A. J. Hudspeth, eds. *Principles of Neural Science*, 5th ed. (McGraw-Hill, 2013), 1260.

35. Yukihiro Hara, "Brain Plasticity and Rehabilitation in Stroke Patients," *Journal of Nippon Medical School* 82, no. 1 (2015): 4–13.

36. Sue Himmelrich, "Statement Apologizing to Santa Monica's African American Residents and Their Descendants," City of Santa Monica, November 16, 2022, https://www.santamonica.gov/blog/statement-apologizing-to-santa-monica-s-african-american-residents-and-their-descendants.

37. Joseph McCowan, in discussion with the author, December 21, 2023.

38. Harriet A. Washington, *Medical Apartheid: The Dark History of Medical Experimentation on Black Americans from Colonial Times to the Present* (Doubleday, 2006).

39. "Racial Trauma," Mental Health American, accessed August 4, 2024, https://www.mhanational.org/racial-trauma.

40. David Hartman and Diane Zimberoff, "The Nourishment Barrier: The Shock Response to Toxic Intimacy," *Journal of Heart-Centered Therapies* 15, no. 2 (2012): 3–26.

41. Jhumpa Lahiri, *Unaccustomed Earth* (Alfred A. Knopf Canada, 2009).

42. Emma Bruehlman-Senecal and Ozlem Ayduk, "This Too Shall Pass: Temporal Distance and the Regulation of Emotional Distress," *Journal of Personality and Social Psychology* 108, no. 2 (2015): 356–75.

43. Lisa Held and Alexandra Rutherford, "Can't a Mother Sing the Blues? Postpartum Depression and the Construction of Motherhood in Late Twentieth-Century America," *History of Psychology* 15, no. 2 (2012): 107–23.

44. Richard Schwartz, *No Bad Parts: Healing Trauma and Restoring Wholeness with the Internal Family Systems Model* (Sounds True Adult, 2021).

45. Anita Barrows and Joanna Macy, trans. and eds., *In Praise of Mortality: Selections from Rainer Maria Rilke's "Duino Elegies" and "Sonnets to Orpheus"* (Echo Point, 2022), 13.

46. "Somatic Experiencing® (SE™)," Ergos Institute of Somatic Education, accessed October 16, 2024, https://www.somaticexperiencing.com/somatic-experiencing; Danny Brom, Yaffa Stokar, Cathy Lawi, et al., "Somatic Experiencing for Posttraumatic Stress Disorder: A Randomized Controlled Outcome Study," *Journal of Traumatic Stress* 30, no. 3 (June 2017): 304–12.

47. "SE™ International Training & Events," Somatic Experiencing® International, accessed October 16, 2024, https://traumahealing.org/training/.
48. Edmund J. Bourne, *The Anxiety and Phobia Workbook* (New Harbinger Publications, Inc., 2020).

Chapter 8: Learning How to Talk to Yourself

1. Peter Hulm, "Rilke's Valais: 'I Have This Country in the Blood,'" *Global Geneva*, October 5, 2018, https://global-geneva.com/rilkes-valais-in-switzerland-i-have-this-country-in-the-blood/.
2. Hulm, "Rilke's Valais."
3. Rainer Maria Rilke to Marie von Thurn und Taxis, 25 July 1921, *Valais Through Rilke's Eyes*, Rainer Maria Rilke Foundation, Sierre, Switzerland, Gallery 1/III, Setting the Scene: Checquered Hills.
4. Rainer Maria Rilke to Xaver von Moos, 2 March 1922, *Valais Through Rilke's Eyes*, Rainer Maria Rilke Foundation, Sierre, Switzerland, Room 1/III, Setting the Scene: Majestic Mountains.
5. Rainer Maria Rilke to Elya Maria Nevar, 9 January 1922, *Valais Through Rilke's Eyes*, Rainer Maria Rilke Foundation, Sierre, Switzerland, Room 1/II, Setting the Scene: Biblical Views.
6. *Valais Through Rilke's Eyes*, Rainer Maria Rilke Foundation, Sierre, Switzerland, Gallery 2/II, Discovery of Muzot.
7. *Valais Through Rilke's Eyes*, Rainer Maria Rilke Foundation, Sierre, Switzerland, Gallery 2/I, Rilke's Affinity for Towers.
8. *Valais Through Rilke's Eyes*, Rainer Maria Rilke Foundation, Sierre, Switzerland, Gallery 2/I, Rilke's Affinity for Towers.
9. "A Storm from Heart and Soul," *Valais Through Rilke's Eyes*, Rainer Maria Rilke Foundation, Sierre, Switzerland, Gallery 3/II *Duino Elegies*.
10. Annamarya Scaccia, "Weird Early Pregnancy Symptoms No One Tells You About," Healthline, May 16, 2023, https://www.healthline.com/health/pregnancy/weird-early-symptoms.
11. Mayo Clinic Staff, "Gestational Diabetes," Mayo Clinic, April 9, 2022, https://www.mayoclinic.org/diseases-conditions/gestational-diabetes/symptoms-causes/syc-20355339.
12. Valentina Nicolardi, Luca Simione, Domenico Scaringi, et al. "The Two Arrows of Pain: Mechanisms of Pain Related to Meditation and Mental States of Aversion and Identification," *Mindfulness* 15, no. 4 (2024): 753–74.
13. Kate Alcamo, in discussion with the author, January 3, 2024.
14. Lilian Dindo, Julia R. Van Liew, and Joanna J. Arch, "Acceptance and Commitment Therapy: A Transdiagnostic Behavioral Intervention for Mental Health and Medical Conditions," *Neurotherapeutics* 14, no. 3 (2017): 546–53.

15. Thomas Marra, *Dialectical Behavior Therapy in Private Practice: A Practical and Comprehensive Guide* (New Harbinger, 2005).
16. Brit Murphy, in discussion with the author, January 3, 2024.
17. Andrada D. Neacsiu, Erin F. Ward-Ciesielski, and Marsha M. Linehan, "Emerging Approaches to Counseling Intervention: Dialectical Behavior Therapy," *The Counseling Psychologist* 40, no. 7 (2012): 1003–32.
18. Marina Kerlow, in discussion with the author, January 5, 2024.
19. Sanjana Gupta, "Feeling Anxious? Try the 5-4-3-2-1 Grounding Technique," *Verywell Mind*, April 29, 2024, https://www.verywellmind.com/5-4-3-2-1-grounding-technique-8639390.
20. Valeria McCarroll, in discussion with the author, January 12, 2024.
21. Ethan Kross, *Chatter: The Voice in Our Head, Why It Matters, and How to Harness It* (Crown, 2021), xi.
22. Kross, *Chatter*, xix.
23. Kross, *Chatter*, xii–xiv.
24. Kross, *Chatter*, xiv.
25. Yaacov Trope and Nira Liberman, "Construal-Level Theory of Psychological Distance," *Psychological Review* 117, no. 2 (2010): 440–63; Ethan Kross, in discussion with the author, December 1, 2023.
26. Kross, *Chatter*, 15.
27. Rainer Maria Rilke, *Letters on Life: New Prose Translations*, trans. Ulrich Baer (Random House, 2017), xv.
28. James Pennebaker, in discussion with the author, December 8, 2023.
29. James W. Pennebaker, "Expressive Writing in Psychological Science," *Perspectives on Psychological Science* 13, no. 2 (2018): 226–29.
30. Maury Joseph, in discussion with the author, January 1, 2024.
31. Maury Joseph, "The Discomfort of Not Knowing: How to Live with Unanswered Questions," *GoodTherapy Blog*, July 31, 2017, https://www.goodtherapy.org/blog/discomfort-of-not-knowing-how-to-live-with-unanswered-questions-0731175.
32. Kaitlin Nunamann, "Four Strategies to Soothe Anxious Uncertainty," *Psychology Today*, September 13, 2022, https://www.psychologytoday.com/us/blog/lying-your-therapist/202209/4-strategies-soothe-anxious-uncertainty.
33. Kaitlin Nunamann, in discussion with the author, January 10, 2024.
34. Bessel van der Kolk, *The Body Keeps the Score: Brain, Mind, and Body in the Healing of Trauma* (Penguin Books, 2015).
35. Kristin D. Neff, "Self-Compassion: Theory, Method, Research, and Intervention," *Annual Review of Psychology* 74, no. 1 (2023): 193–218.

Chapter 9: Holding On and Letting Go

1. Ralph Freedman, *Life of a Poet: Rainer Maria Rilke* (Northwestern Univ. Press, 1996), 531–32.
2. Freedman, *Life of a Poet*, 535.
3. Freedman, *Life of a Poet*, 541–48.
4. Rainer Maria Rilke and Lou Andreas-Salomé, *Rilke and Andreas-Salomé: A Love Story in Letters*, trans. Edward Snow and Michael Winkler (W. W. Norton, 2016), 360.
5. Ulrich Baer, in discussion with the author, January 8, 2024, and February 16, 2024.
6. Ulrich Baer, ed., *110 Stories: New York Writes After September 11* (NYU Press, 2002).
7. Evan Imber-Black, "Rituals in the Time of COVID-19: Imagination, Responsiveness, and the Human Spirit," *Family Process* 59, no. 3 (September 2020): 912–21.
8. Jamie Dettmer, "The World Shuts Down—but Italy Sings," *VOA*, March 13, 2020, https://www.voanews.com/a/science-health_coronavirus-outbreak_world-shuts-down-italy-sings/6185782.html.
9. Lorri Hinnant, "Europeans Sing Health Workers' Praises Nightly from Windows," Associated Press, March 19, 2020, https://apnews.com/article/virus-outbreak-health-spain-ap-top-news-paris-63bb9038a3fc719cb1370cd8ce4773b3.
10. "Spanish Police Sing to Families During Coronavirus Lockdown in Mallorca," posted March 23, 2020, by Guardian News, YouTube, 34 sec., https://www.youtube.com/watch?v=mEpkUawiLKA.
11. Gary Hardcastle, "Every Night, New York City Salutes Its Health Care Workers," NPR, April 10, 2020, https://www.npr.org/2020/04/10/832131816/every-night-new-york-city-salutes-its-health-care-workers.
12. Cristine H. Legare, "Why Rituals Are Important Pandemic Survival Tools," BBC, February 4, 2021, https://www.bbc.com/worklife/article/20210203-why-rituals-are-important-pandemic-survival-tools.
13. Casper ter Kuile, *The Power of Ritual: Turning Everyday Activities into Soulful Practices* (HarperOne, 2020), 25.
14. ter Kuile, *Power of Ritual*, 25.
15. ter Kuile, *Power of Ritual*, 26.
16. Michael Norton, *The Ritual Effect: From Habit to Ritual, Harness the Surprising Power of Everyday Actions* (Simon & Schuster, 2024).
17. Michael "Mike" Norton, in discussion with the author, December 16, 2023.
18. Nicholas Hobson, Juliana Schroeder, Jane Risen, Dimitris Xygalatas, and Michael Inzlicht, "The Psychology of Rituals: An Integrative Review and

Process-Based Framework," *Personality and Social Psychology Review* 22, no. 3 (2018): 260–84.

19. Martin Lang, Jan Krátky, and Dimitris Xygalatas, "The Role of Ritual Behaviour in Anxiety Reduction: An Investigation of Marathi Religious Practices in Mauritius," *Philosophical Transactions B* 375, no. 1805 (2020): 20190431.

20. Casper ter Kuile, in discussion with the author, July 7, 2023.

21. Ori Tavor, "Embodying the Dead: Ritual as Preventative Therapy in Chinese Ancestor Worship and Funerary Practices," *Journal of Ritual Studies* 34, no. 1 (2020): 31–42.

22. Curie Virág, "Rituals Create Community by Translating Our Love into Action," *Psyche*, July 28, 2021, https://psyche.co/ideas/rituals-create-community-by-translating-our-love-into-action; Curie Virág, in discussion with the author, February 13, 2024.

23. Baer, discussion.

24. Virág, discussion.

25. S. Josephine Baker, *Fighting for Life* (New York Review Books, 2013), 1.

26. Baker, *Fighting for Life*, 20–21.

27. Baker, *Fighting for Life*, 21.

28. Baker, *Fighting for Life*, 21.

29. Baker, *Fighting for Life*, vi–x.

30. Baker, *Fighting for Life*, x.

31. Baker, *Fighting for Life*, 49–50.

32. Baker, *Fighting for Life*, 73–75.

33. Baker, *Fighting for Life*, 182.

34. Joanna Scutts, *Hotbed: Bohemian Greenwich Village and the Secret Club That Sparked Modern Feminism* (Basic Books, 2022), 4.

35. Judith Schwarz, *Radical Feminists of Heterodoxy: Greenwich Village, 1912–1940* (New Victoria, 1986), 23, 40.

36. Schwarz, *Radical Feminists of Heterodoxy*, ii; Scutts, *Hotbed*, 3–4.

37. John Fabian Witt, "Internationalism and the Dilemmas of Strategic Patriotism," *Tulsa Law Review* 41 (2005): 787.

38. Adam Hochschild, *Rebel Cinderella: From Rags to Riches to Radical, the Epic Journey of Rose Pastor Stokes* (Houghton Mifflin Harcourt, 2020).

39. Mabel Dodge Luhan, *Intimate Memories: Movers and Shakers*, vol. III (Harcourt, Brace, 1936), 143.

40. Scutts, *Hotbed*, 9, 14–15.

41. Schwarz, *Radical Feminists of Heterodoxy*, 18.

42. National Center for Health Statistics, *100 Years of Marriage and Divorce Statistics, United States 1867–1967* (US Government Printing Office, 1973).
43. Scutts, *Hotbed*, 8, 15.
44. Schwarz, *Radical Feminists of Heterodoxy*, 18.
45. Schwarz, *Radical Feminists of Heterodoxy*, 23.
46. Clair Chiaratti, Marie Madison Denayer, and Melanie Hanly, "Leta Stetter Hollingworth," in *Open History of Psychology: The Lives and Contributions of Marginalized Psychology Pioneers*, eds. Alison E. Kelly and Brittany N. Avila (Press Books, 2023), https://pressbooks.pub/openhistoryofpsychology/; Ann G. Klein, *A Forgotten Voice: A Biography of Leta Stetter Hollingsworth* (Great Potential Press, 2002); Kate E. Wittenstein, *The Heterodoxy Club and American Feminism, 1912–1930* (PhD diss., Boston Univ., 1989), 85–86, 101–2.
47. Rheta Childe Dorr, *A Woman of Fifty* (Funk & Wagnalls, 1924), 277.
48. Judith Schwarz, "The Archivist's Balancing Act: Helping Researchers While Protecting Individual Privacy," *The Journal of American History* 79, no. 1 (1992): 179–89, 182.
49. Schwarz, *Radical Feminists of Heterodoxy*, 7.
50. Schwarz, *Radical Feminists of Heterodoxy*, 1–2.
51. Scutts, *Hotbed*, 3.
52. Joanna Scutts, in discussion with the author, June 27, 2023.
53. Christoph Irmscher, *Max Eastman: A Life* (Yale Univ. Press, 2017).
54. Sarah Schnitker, in discussion with the author, February 13, 2024.
55. Parker Palmer, in discussion with the author, January 23, 2024.
56. Joan Sutherland, in discussion with the author, March 9, 2024.
57. Annie Duke, *Quit: The Power of Knowing When to Walk Away* (Portfolio, 2022), 15.
58. Schnitker, discussion.
59. F. LeRon Shults and Steven J. Sandage, *Transforming Spirituality: Integrating Theology and Psychology* (Baker Academic, 2006).
60. Vineet Chander, in discussion with the author, August 15, 2024.
61. Baer, discussion.
62. Norton, discussion.
63. Rainer Maria Rilke and Franz Xaver Kappus, *Letters to a Young Poet: With the Letters to Rilke from the "Young Poet,"* trans. Damion Searls (Liveright, 2020), 27.

Epilogue: Where to Find Courage?

1. Rainer Maria Rilke and Lou Andreas-Salomé, *Rilke and Andreas-Salomé: A Love Story in Letters*, trans. Edward Snow and Michael Winkler (W. W. Norton, 2016), 360.

2. Ralph Freedman, *Life of a Poet: Rainer Maria Rilke* (Northwestern Univ. Press, 1996), 473.
3. Freedman, *Life of a Poet*, 492–93.
4. Rainer Maria Rilke to Nanny Qunderly-Volkart, 4 July 1921, *Valais Through Rilke's Eyes*, Rainer Maria Rilke Foundation, Sierre, Switzerland, Gallery 2/III, Bucolic Living.
5. Rainer Mara Rilke to Elya Maria Nevar, 9 January 1922, *Valais Through Rilke's Eyes*, Rainer Maria Rilke Foundation, Sierre, Switzerland, Room 1/II, Setting the Scene: Biblical Views.
6. *Valais Through Rilke's Eyes*, Rainer Maria Rilke Foundation, Sierre, Switzerland, Room 2/II, Discovery of Muzot.
7. Rainer Maria Rilke, *The Poetry of Rilke*, trans. Edward Snow, introduction by Adam Zagajewski (North Point Press, 2011), ix.
8. Rilke, *The Poetry of Rilke*, ix.
9. Rilke and Betz, *Rilke in Paris*, 82.
10. Rilke and Betz, *Rilke in Paris*, 83.
11. Rainer Maria Rilke, *The Inner Sky: Poems, Notes, Dreams*, trans. Damion Searls (David R. Godine, 2010), 45.

Appendix

1. Judson Brewer, *Unwinding Anxiety: New Science Shows How to Break the Cycles of Worry and Fear to Heal Your Mind* (Avery, 2021), xiii, 39, 49.
2. Brewer, *Unwinding Anxiety*, 72.
3. Brewer, *Unwinding Anxiety*, 172, 181–82.
4. Brewer, *Unwinding Anxiety*, 69.
5. Scott Shigeoka, *Seek: How Curiosity Can Transform Your Life and Change the World* (Balance, 2023), 21.
6. Shigeoka, *Seek*, 108.
7. Shigeoka, *Seek*, 113.
8. Brewer, *Unwinding Anxiety*, 141.

ABOUT THE AUTHOR

ELIZABETH WEINGARTEN is a journalist and applied behavioral scientist who works at the intersection of science and storytelling. She has worked on the editorial staffs of *The Atlantic*, *Slate*, and *Qatar Today*, and was managing editor of *Behavioral Scientist*. Her writing has appeared in publications including *The Atlantic*, *Slate*, *CNN*, *Financial Times*, *Harvard Business Review*, and *TIME*. She has led research programs at the think tank New America, the consultancy ideas42, and the tech companies Torch and Udemy. She lives in Northern California with her husband and son. You can learn more about her work by visiting www.elizabethweingarten.com.